Dear Parent,

Welcome to a new school year! This letter is to introduce you to **A Reason For**© Science.

A Reason For© Science teaches basic Life, Earth, and Physical Science through fun, hands-on activities. Each lesson is tied directly to the **National Science Education Content Standards** and uses an inquiry-based approach designed to enhance learning.

Today's increasingly complex world requires a clear understanding of science and technology. Our future prosperity depends on helping children rediscover the challenges, excitement, and joy of science — especially in the context of Scripture values. Thus, one of the primary goals of **A Reason For**© Science is to make science not only meaningful, but also FUN!

Fun, Flexible Format

Instead of a hardback textbook filled with "facts" to memorize, your child will be working in an interactive worktext designed to develop critical-thinking skills. Students start each week with a hands-on activity demonstrating a key science concept. This is followed by group discussion, journaling, and a series of thought-provoking questions. Lessons conclude with a summary of key concepts and a related "object lesson" from Scripture.

Safety Issues

The hands-on nature of **A Reason For**© Science means your child will be working with age-appropriate materials. (For instance, the "acids" we use are actually dilute forms comparable to typical household chemicals.) Like a field trip or gym class, these science activities usually require simple safety precautions.

But for instructional reasons, all materials in **A Reason For**© Science are treated as hazardous! This encourages students to develop good safety habits for use in later years. (If you have further questions about safety, your child's teacher has an in-depth safety manual outlining precautions for every lesson.)

Scripture Values

Best of all, **A Reason For**© Science features Scripture values! Every lesson concludes with a Scripture object lesson related to the week's topic. These "Food for Thought" sections encourage students to relate everyday experiences to Scriptural themes, providing a positive way to integrate faith and learning.

Here's to an exciting year exploring God's world!

Dave & Rozann Seela
Authors, **A Reason For**© Science

A reason for Science

Hands-On Activities   With Scripture Values

LEVEL F

STUDENT WORKTEXT

ISBN #1-58938-145-9

Published by The Concerned Group, Inc.
700 East Granite • PO Box 1000 • Siloam Springs, AR 72761

Authors	**Dave & Rozann Seela**
Publisher	**Russ L. Potter, II**
Senior Editor	**Bill Morelan**
Project Coordinator	**Rocki Vanatta**
Creative Director	**Daniel Potter**
Proofreader	**Elizabeth Granderson**
Step Illustrations	**Steven Butler**
Character Illustrations	**Josh Ray**
Colorists	**Josh & Aimee Ray**

Printed on recycled paper in the United States

For more information about **A Reason For**® curricula,
write to the address above, call, or visit our website.

www.areasonfor.com
800.447.4332

A Reason For© Science makes science FUN! Your school year will be filled with hands-on activities, colorful discovery sheets, and lots of discussion and exploration. You'll discover many exciting new things as you explore God's world!

Although "almost everything relates to everything else" in some way or another, scientists usually divide science into three broad areas for study: **life**, **earth,** and **physical** science. The sections in your worktext are based on these categories.

Colorful icons are used to help you identify each section. An **ant** represents **Life Science** lessons. A **globe** stands for **Earth Science** lessons. An **atom** introduces **Physical Science (Energy/Matter)** lessons. And a **hammer** represents the **Physical Science (Forces)** lessons.

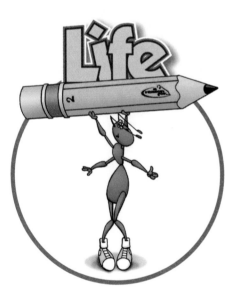

Life Science

Life Science is the study of **living things.** In the Life Science section of your **A Reason For©** Science worktext, you'll explore different kinds of living things. You'll learn about their characteristics (how they're alike or different). You'll discover how scientists classify (label and sort) living things. You'll even learn more about your own body and how it works!

Earth Science

Earth Science is the study of **earth** and **sky**. In the Earth Science section of your **A Reason For©** Science worktext, you'll explore the structure of our planet (rocks, crystals, volcanos), the atmosphere (air, clouds), and related systems (water cycles, air pressure, weather). Grades 7 and 8 reach out even further with a look at the solar system and stars.

Physical Science

Physical Science is the study of **energy, matter,** and related **forces.** Since the physical part of science has a big effect on your daily life, it's divided into two major sections:

Energy and Matter

In this section of your **A Reason For©** Science worktext, you'll learn about the different **states** and unique **properties** of **matter**. You'll discover new things about light and sound. You'll explore physical and chemical reactions. Some grades will explore related concepts like circuits, currents, and convection.

Forces

In this section of your **A Reason For©** Science worktext, you'll discover how "**push** and **pull**" form the basis for all physical movement. You'll explore simple machines (levers, pulleys). You'll work with Newton's laws of motion. You'll even learn to understand concepts like torque, inertia, and buoyancy.

Safety First!

Before you begin, be sure to read about "Peat" the safety worm. Peat's job is to warn you whenever there's a potential hazard around. (Whenever you see Peat and his warning sign, STOP and wait for further instructions from your teacher!)

LEARN TO BE SAFE!

Exploring God's world of science often requires using equipment or materials that can injure you if they're not handled correctly. Also, many accidents occur when people hurry, are careless, or ignore safety rules.

It's your responsibility to know and observe the rules and to use care and caution as you work. Just like when you're on the playground, horseplay or ignoring safety rules can be dangerous. Don't let an accident happen to you!

Meet "Peat" the Safety Worm!

Peat's job is to warn you whenever there's a potential hazard around. Whenever you see Peat and his warning sign, **STOP** and wait for further instructions from your teacher!

Peat's sign helps you know what kind of hazard is present. Before beginning each activity, your teacher will discuss this hazard in detail and review the safety rules that apply.

means this activity requires **PROTECTIVE GEAR.**

Usually **gloves** or **goggles** (or both) are required. Goggles protect your eyes from things like flying debris or splashing liquids. Gloves protect your hands from things like heat, broken glass, or corrosive chemicals.

means there is a BURN HAZARD. There are three common burn hazards.

"Open Flame" indicates the presence of fire (often matches or a candle). "Thermal Burn" means objects may be too hot to touch. "Corrosion" indicates a chemical substance is present.

means there is a POISON HAZARD.

There are three common poison hazards. "Skin Contact" indicates a substance that should not touch skin. "Vapor" indicates fumes that should not be inhaled. "Hygiene" indicates the presence of materials that may contain germs.

indicates OTHER HAZARDS.

There are three additional hazards that require caution. **"Breakage"** indicates the presence of fragile substances (like glass). **"Slipping"** indicates liquids that might spill on the floor. **"Sharp Objects"** indicates the presence of tools with sharp edges or points.

Play It Safe!

Exploring God's world with **A Reason For©** Science can be great fun, but remember — play it safe! Observe all the safety rules, handle equipment and materials carefully, and always be cautious and alert.

And don't forget Peat the safety worm! Whenever you see Peat and his warning sign, STOP and wait for further instructions from your teacher.

Life Science

Life Science is the study of **living things.** In this section, you'll explore different kinds of living things. You'll learn about their characteristics (how they're alike or different). You'll discover how scientists classify (label and sort) living things. You'll even learn more about your own body and how it works!

PEANUT POWER

FOCUS Seed Structure

OBJECTIVE To explore the structure and purpose of seeds

OVERVIEW There are many different kinds of seeds, but they all have things in common. In this activity, we'll dissect a seed to find out how it works.

WHAT TO DO

STEP 1

Hold a peanut in your fingers. **Look** closely at both ends and **describe** any differences. **Observe** the shell, especially its shape and texture. Now gently **crack** the shell. **Open** it lengthwise along the seams. **Record** (and **draw**) your observations.

STEP 2

Carefully **lift out** one seed (the nut). **Observe** the thin red covering. **Remove** the covering if it hasn't already fallen off. **Examine** the seed closely and **make notes** about its shape and texture.

STEP 3

Use your thumbnail to carefully **split** the seed lengthwise along its seam. **Find** the bump between the halves (usually on one end). **Predict** what this might be. **Examine** this part closely, then **draw** the seed in your journal.

STEP 4

Discard the remains of your peanut project as directed by your teacher. Now **review** each step in this activity. **Share** and **compare** observations with your research team.

WHAT HAPPENED?

Seeds are the foundation for the next generation of **plants**. Locked inside every seed is an **embryo** (baby plant) covered by a protective **seed coat**, plus **food** for the embryo as it starts to grow. This growth process (**germination**) begins when the correct amount of **moisture** and **heat** becomes available.

First, the seed absorbs **water**. As the seed swells, it splits the seed coat so that the embryo can grow out. Then the **roots** begin to grow down in a **behavior** called **geotropism**. "Tropos" means "turn" and "geo" means "Earth," so geotropism indicates a turning toward Earth. At the same time, the seed's top grows upward in a behavior called **phototropism**. "Photo" means "light," so phototropism indicates turning toward **light energy**. The roots keep digging deeper, and the leaves keep pushing upward — and you have a new peanut plant!

WHAT WE LEARNED

 What is the purpose of a peanut's shell? Based on your observations, why do you think the peanut's shell is well suited for this purpose?

_____

_____

_____

_____

_____

_____

_____

 Describe the shape, color, and texture of the seed you removed in Step 2. What is the purpose of the red skin covering?

_____

_____

_____

_____

_____

_____

3 What was the small, curved "bump" you found in Step 3?
Describe the section of the seed in detail.

_____

_____

_____

_____

_____

_____

_____

4 What is the purpose of the material surrounding the bump?
How does this help the new plant?

_____

_____

_____

_____

_____

_____

_____

5 What are the plant behaviors that make roots grow down and leaves
grow up? Why are they important to a plant's survival?

_____

_____

_____

_____

_____

_____

_____

CONCLUSION

Seeds contain an embryo and food to start its growth. Seeds not only create new plants, but are also an important food source for people and animals.

FOOD FOR THOUGHT

Genesis 8:22 Seeds are a vital part of the life cycle on Earth. Seeds grow into plants, plants grow and are harvested for food, and more seeds are planted to continue the cycle. God has promised that this cycle will continue as long as the Earth remains.

Yet even though God maintains this great cycle of life to provide for our needs, we often take natural things (like food cycles, sunrise, and rain) for granted. Be sure to take time this week to thank God for these wonderful blessings!

JOURNAL   My Science Notes

_____

_____

_____

_____

_____

_____

_____

_____

_____

_____

_____

_____

_____

_____

_____

_____

_____

_____

_____

_____

SCATTERING SEEDS

FOCUS Seed Dispersal

OBJECTIVE To explore how seeds are scattered around

OVERVIEW Hook and loop tape (like Velcro®) is pretty handy stuff! Where did inventor George de Mestral get the idea for this amazing product? How does it work? In this activity, we'll find out!

WHAT TO DO

STEP 1

Pick up the cocklebur seed. **Look** at it closely with the magnifying lens. **Make notes** in your journal describing what you see.

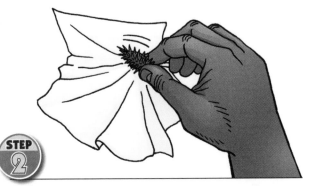

STEP 2

Feel the hooks gently with your fingertip. (Be careful! Some hooks are like thorns!) Gently **rub** the seed against a piece of cloth. **Make notes** about what happens. Now **pull** the seed away from the cloth. **Record** the results.

STEP 3

Using the magnifying lens, **look** closely at the two pieces of Velcro®. **Make notes** describing how the two pieces are different. Now **push** the pieces together, then **pull** them apart. **Make notes** about what happens.

STEP 4

Review each step in this activity. **Make notes** about how the seeds and the Velcro® were similar. **Share** and **compare** observations with your research team.

WHAT HAPPENED?

Plants must produce lots of **seeds** because many end up in locations where they can't grow. Not only that, but if all of a plant's seeds landed in the same spot, there wouldn't be enough **water**, **sun**, or **soil** to go around.

God gave many plants unique ways to scatter (**disperse**) their seeds. Some seeds are light and blow away on the **wind**. Some float on the water, drifting great distances down rivers and streams. Some are **hitchhikers**, using little hooks to hitch rides on passing animals. Inventor George de Mestral wondered how such seeds attached themselves to his dog. He looked closer, and as a result he eventually developed a man-made version to create a quick, handy fastener!

WHAT WE LEARNED

 Describe what you discovered in your examination of the cocklebur in Step 1.

_____

_____

_____

_____

_____

_____

_____

 Describe the interaction between the cocklebur and the cloth in Step 2. Why did this happen?

_____

_____

_____

_____

_____

_____

_____

3 Describe the difference between the two strips of Velcro®. Which strip is similar to a cocklebur? Which strip is similar to the cloth?

_____

_____

_____

_____

_____

_____

4 Compare the interaction between the cocklebur and cloth in Step 2 with the interaction between the strips of Velcro in Step 3. How were they similar? How were they different?

_____

_____

_____

_____

_____

_____

5 What advantage do the hooks give a cocklebur plant? Why do plants want to scatter their seeds as far away as possible?

_____

_____

_____

_____

_____

_____

CONCLUSION

Plants scatter their seeds in a variety of ways. The dispersal of seeds avoids over-crowding, and helps insure the survival of plants.

FOOD FOR THOUGHT

Proverbs 18:24 Cockleburs and similar seeds can really stick to you, even if you're bouncing around or the road gets really rough! Once they're attached, you can be sure they'll always try to stay right there with you.

This Scripture reminds us there are two kinds of friends — someone who pretends to be friendly, and someone who sticks close to you no matter what happens. Remember, God is your greatest friend! Wherever life may take you, he can always be there by your side. As long as you trust him, God will never abandon you. Who would you rather spend more time with — a pretend friend, or a real friend?

 **My Science Notes**

_____

_____

_____

_____

_____

_____

_____

_____

_____

_____

_____

_____

_____

_____

_____

_____

_____

_____

_____

_____

LEACHING LEAVES

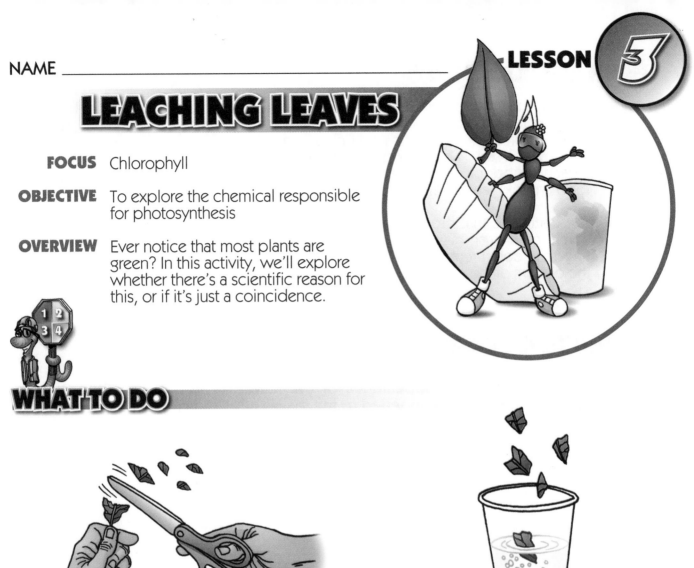

FOCUS   Chlorophyll

OBJECTIVE   To explore the chemical responsible for photosynthesis

OVERVIEW   Ever notice that most plants are green? In this activity, we'll explore whether there's a scientific reason for this, or if it's just a coincidence.

WHAT TO DO

STEP 1

Join your teacher on a research expedition to collect leaves. You must **find** a large, green leaf for this activity. **Return** to your work surface and **cut** the leaf into tiny pieces. **Observe** the pieces and **make notes** in your journal.

STEP 2

Pour one ounce of fingernail polish remover into a paper cup. **Drop** the leaf pieces into the cup, then **swirl** it gently. **Move** the cup to the location your teacher indicates and let it set overnight.

STEP 3

Cut a one-inch strip of coffee filter or paper towel. **Predict** what might happen if you put one end in the liquid. Now carefully **lower** the filter into the cup. Make sure it makes good contact with the liquid. **Record** the results.

STEP 4

Wait 60 seconds, then **look** at the strip closely. **Make notes** about what you see, including any colors other than green. **Share** and **compare** observations with your research team.

 WHAT HAPPENED?

The fingernail polish remover **dissolved** an important plant chemical (**chlorophyll**) that was then **absorbed** by the filter paper. The presence of chlorophyll causes the green color in **plants**. Plants need chlorophyll for a vital **chemical reaction** called **photosynthesis**. Photosynthesis combines **carbon dioxide** from the **air** with **water**, creating **food** for the plant and releasing **oxygen**. Human survival depends on the food and oxygen plants produce!

The chemicals that produce colors in living things are called **pigments**. If you saw colors other than green on your filter strip, other pigments were present. As the fingernail polish remover was absorbed, large pigment particles were trapped near the bottom of the paper, while smaller pigment particles moved further up the strip.

 WHAT WE LEARNED

 Compare the appearance of the leaves in Step 1 with Step 3. What color change happened to the leaves after soaking? What color change occurred in the solution?

_____

_____

_____

_____

_____

_____

 What did you predict in Step 3? How did your prediction reflect what actually happened?

_____

_____

_____

_____

_____

_____

What is the name of the chemical that makes plants green? Why is this chemical important to plants?

_____

_____

_____

_____

_____

_____

What is the name of the chemical reaction that takes place in plants? What two products does this produce? Why is this important for human survival?

_____

_____

_____

_____

_____

_____

_____

Notice the colors on your filter paper. Which color has the largest pigment particles? Which has the smallest? How do you know this?

_____

_____

_____

_____

_____

_____

CONCLUSION

The chemical chlorophyll causes the green color in plants. Plants need chlorophyll for the photosynthesis process that creates food and releases oxygen. These plant traits are essential for human survival.

FOOD FOR THOUGHT

1 Peter 3:3, 4 Many plants seem to be just one color — green! But in this activity, we discovered pigments that weren't visible on the outside of the plant. These beautiful colors became visible when we took the extra effort to find them.

This Scripture reminds us that it's what's inside that makes someone beautiful. Just as the leaves contained hidden pigments, so there are often precious things hidden inside people's hearts. Instead of focusing on mere outward beauty (clothes, hair, etc.), we should learn to look for a beautiful spirit in others. By keeping close to God, we can learn to cultivate that kind of inward beauty ourselves!

JOURNAL   My Science Notes

_____

_____

_____

_____

_____

_____

_____

_____

_____

_____

_____

_____

_____

_____

_____

_____

_____

_____

_____

_____

LESSON 4

BANANA BAG

FOCUS Decomposition

OBJECTIVE To explore how organisms decompose after death

OVERVIEW Plants use sunlight to store energy, but do they need anything else to survive? In this activity, we'll explore the role decomposition plays in a plant's life cycle.

WHAT TO DO

STEP 1

Start with two plastic bags. **Label** Bag 1: "DO NOT OPEN! Active Ingredient = Yeast." **Label** Bag 2: "DO NOT OPEN! Active Ingredient = None." Now carefully **slice** a banana into several pieces.

STEP 2

Wash your hands thoroughly, then **add** three banana slices to Bag 1. Carefully **sprinkle** a teaspoon of yeast over the banana slices. Now **seal** the bag tightly. **Predict** what might happen in this bag over time.

STEP 3

Wash your hands thoroughly, then **add** three banana slices to Bag 2. Make absolutely certain no yeast has touched the slices! Now **seal** the bag tightly. **Predict** what might happen in this bag over time.

STEP 4

Check both bags every day. **Make notes** about what you see. (**Include** the time and date with each entry.) At the end of the week, **share** and **compare** observations with your research team. **Dispose** of the bags as directed by your teacher.

WHAT HAPPENED?

Scientists call most material produced by living creatures **organic**. Organic **matter** is constantly being attacked by **bacteria**, **fungi** (like yeast), and other **microbes**, whichattempt to break it down through a process called **decomposition**. The microbes that cause decomposition are everywhere. That's why it's so important to always wash your hands and the utensils you use!

As the decomposition process takes place, **nutrients** (plant foods) are released. **Plants** need these nutrients to grow and develop. Decomposition also releases by-products. As the yeast munched on the sugar in the bananas, it gave off **carbon dioxide gas** — which caused the bag to swell. Decomposition is very important because it allows plants to **recycle** chemicals that dead creatures no longer need and to provide essential nutrients.

WHAT WE LEARNED

 Why wash your hands before beginning? What might happen to the bags' contents if you didn't? Why wash your hands when finished? What might happen to you if you didn't?

_____

_____

_____

_____

_____

_____

_____

 What did you predict in Step 2? What did you predict in Step 3? How did these predictions reflect what actually happened?

_____

_____

_____

_____

_____

_____

_____

3 What are microbes, and what role do they play in a plant's life cycle? Why is decomposition important to all living things?

_____

_____

_____

_____

_____

_____

4 Compare the two bags in Step 4. How were they similar? How were they different?

_____

_____

_____

_____

_____

_____

5 Which bag decomposed the banana slices the best? Why? How was it different from the other bag?

_____

_____

_____

_____

_____

_____

CONCLUSION

Microbes cause all organic matter to decompose. Decomposition releases essential nutrients so they can be reused. Decomposition is an important part of the life cycle of all living things.

FOOD FOR THOUGHT

John 14:26 The banana plant that produced your banana depended on nutrients released by microbes through decomposition. These nutrients continually recycle, moving through all sorts of living things. Without the constant work of microbes, these nourishing elements could not return to the soil for plants and animals to reuse.

This Scripture reminds us that God has a special helper who is constantly working, too! Like those tireless microbes, the Holy Spirit continually works to nourish our lives, reminding us of the things that Jesus said. This comforting process feeds our souls, releasing God's love in our lives and helping us share that love with others.

JOURNAL   My Science Notes

SAVING SALT

FOCUS   Preservatives

OBJECTIVE   To explore how preservatives affect decomposition

OVERVIEW   Microbes cause food to spoil (decompose) if it's not protected. How can you stop this process without a refrigerator? In this activity, we'll explore one alternative.

WHAT TO DO

STEP 1

Start with two plastic bags. **Label** Bag 1: "DO NOT OPEN! Active Ingredient = Salt." **Label** Bag 2: "DO NOT OPEN! Active Ingredient = None." Now carefully **slice** a banana into several pieces.

STEP 2

Wash your hands thoroughly, then **add** three banana slices to Bag 1. Carefully **sprinkle** two teaspoons of salt over the banana slices. Now **seal** the bag tightly. **Predict** what might happen in this bag over time.

STEP 3

Wash your hands thoroughly, then **add** three banana slices to Bag 2. Make absolutely certain no salt has touched the slices! Now **seal** the bag tightly. **Predict** what might happen in this bag over time.

STEP 4

Check both bags every day. **Make notes** about what you see. (**Include** the time and date with each entry). At the end of the week, **share** and **compare** observations with your research team. **Dispose** of the bags as directed by your teacher.

WHAT HAPPENED?

Plants use **nutrients**, they **grow**, they **die**, they **decompose**, and they change back into nutrients to start the **life cycle** over again. While humans are dependent on this cycle for survival, sometimes we prefer to change the pace a bit, allowing us to store food for later use. **Preservatives** (like salt) slow or stop the decomposition process by interrupting the action of **microbes**. Salt does this by removing essential **water**, while other types of preservatives use different means.

Preventing **spoilage** means more (and cheaper) food for humans. Although God created the Earth to supply an abundance of good things to eat, food supplies do have a limit. Humans **compete** with everything from microbes to mice for this food. Preservatives are one way to keep food from being wasted, and to keep it available when it's out of season.

WHAT WE LEARNED

 Why are labels (like those you applied to the bags in Step 1) so important in Science? What might happen without labels?

_____

_____

_____

_____

_____

_____

 What did you predict in Step 2? What did you predict in Step 3? How did these predictions reflect what actually happened?

_____

_____

_____

_____

_____

_____

3 What is the role of a preservative? How does it keep food from spoiling?

_____

_____

_____

_____

_____

_____

4 Compare the two bags in Step 4. How were they similar? How were they different?

_____

_____

_____

_____

_____

_____

5 Which bag preserved the banana slices the best? Why? How was it different from the other bag?

_____

_____

_____

_____

_____

_____

CONCLUSION

Although microbes cause all organic matter to decompose, preservatives can help slow or stop this process.

FOOD FOR THOUGHT

Mark 9:50 You've seen that salt can keep food from spoiling. In ancient times, before refrigeration and modern preservatives, salt was even more important. Often salt was the difference between having food and not having food — the difference between life and death!

In this Scripture, Jesus compares his followers to salt. We can truly save lives by sharing God's love with others. But if we're not refilled with God's love each day, we soon become like salt that isn't "salty" anymore. We have nothing left to share. Don't become worthless salt! Take time for God every day.

JOURNAL   My Science Notes

NAME _____

SCIENTIFIC SORTING

FOCUS Classification

OBJECTIVE To explore grouping objects by characteristics

OVERVIEW We can group students by grade, by height, or by age. But how do you group beans? In this activity, we'll explore how scientists use a process called classification.

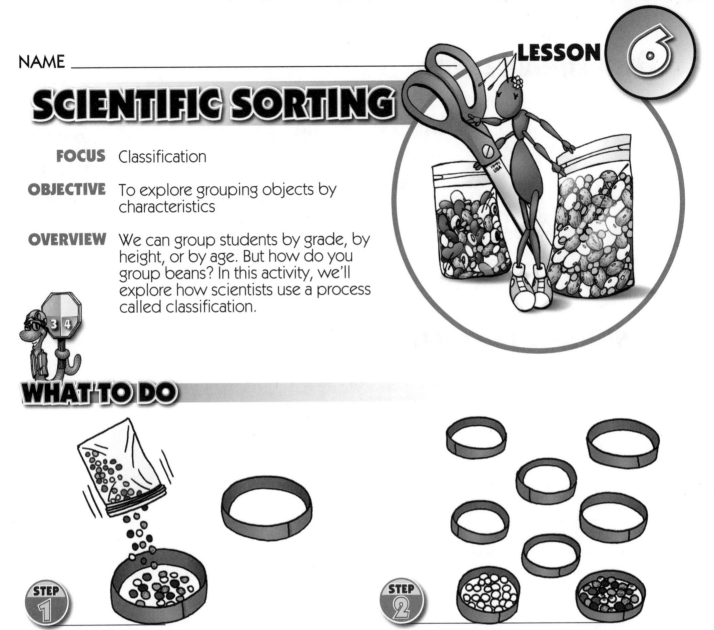

WHAT TO DO

STEP 1

Examine the "few beans" and "many beans" bags. Make notes comparing the contents. Now cut a sheet of construction paper into one inch strips lengthwise. Tape the ends together to make loops. Pour the "few beans" bag into one loop.

STEP 2

In your journal, write a yes/no question about how the beans look. (Example: "Are any of the beans white?") Sort the beans into two groups based on the answer. Now look at the two groups and write another question. Sort the beans into three groups.

STEP 3

Continue sorting beans and writing questions until you fill all the loops. Make notes and drawings about the groups. When everyone on your team is finished, lift the loops, scoop up the beans, and place them back in the bag.

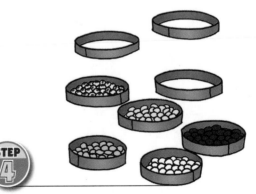

STEP 4

Repeat Steps 1, 2, and 3 using the "many beans" bag. Make notes about these new groups. Discuss how the "many beans" process was different. Put the beans back in the bag, then share and compare observations with your research team.

WHAT HAPPENED?

Sorting **organisms** based on their **characteristics** is called **classification** (also known as **taxonomy**). Scientists use classification to name and divide living things into logical **groups**. This allows scientists from all over the world to know that they're studying and discussing the same kind of creature!

For example, the **scientific name** for humans is *Homo sapiens*. This is a Latin phrase, roughly translated as "thinking man." Scientists always use Latin because it's a "dead" language — it never changes. Since no present-day country speaks it, no new words are being added or old words being given new meanings.

Here's a grammar note: When you're using scientific terms, the entire name is always italicized, but only the first word is capitalized.

WHAT WE LEARNED

 Compare the two bags from Step 1. How were they similar? How were they different?

_____

_____

_____

_____

_____

_____

_____

 What was your first question in step 2? How did it affect your sorting?

_____

_____

_____

_____

_____

_____

_____

3 What were the rest of your sorting questions? How did they affect the sorting process?

_____

_____

_____

_____

_____

_____

_____

4 What is classification based on? What does it allow scientists to do? Why is this important?

_____

_____

_____

_____

_____

_____

_____

5 Compare your final groups with other teams' groups. How were their groups similar? How were they different? What could have happened if you'd all used identical questions?

_____

_____

_____

_____

_____

_____

_____

CONCLUSION

Organisms can be classified into groups based on their characteristics. Scientists use this system to make certain they are observing or discussing the same thing.

FOOD FOR THOUGHT

Luke 14:26 In this activity, you sorted beans into groups based on their characteristics. Once you really understood the question, it was simply a matter of saying "yes" or "no." The correct answer was very clear.

In this Scripture, Jesus asks a very clear question — "Do you love me more than anything else?" If you truly want to be a follower of Jesus, you must first decide on your answer. But just like sorting beans, the more time you spend with God, the easier the correct answer becomes. Why not say "yes" to Jesus, and let God be the master of your life?

JOURNAL   My Science Notes

_____
_____
_____
_____
_____
_____
_____
_____
_____
_____
_____
_____
_____
_____
_____
_____
_____
_____
_____
_____
_____

TERRITORIAL TOOTHPICKS

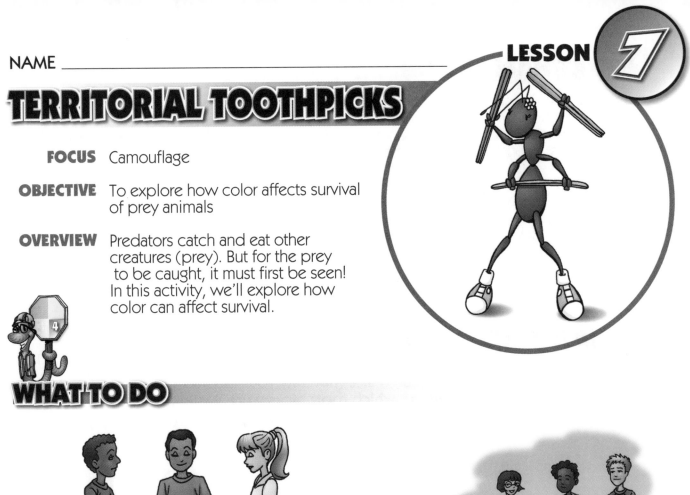

FOCUS   Camouflage

OBJECTIVE   To explore how color affects survival of prey animals

OVERVIEW   Predators catch and eat other creatures (prey). But for the prey to be caught, it must first be seen! In this activity, we'll explore how color can affect survival.

WHAT TO DO

STEP 1

Follow your teacher to the research site. When it's your team's turn, **enter** Toothpick Territory and **pick up** three toothpicks. **Stop** when you have three toothpicks or when time is called. **Note** the amount and color of toothpicks your team gathered.

STEP 2

Return found toothpicks to the teacher. Any team member who didn't find three toothpicks must join the "starved" hunters. Now **watch** the other teams hunt. **Discuss** which toothpicks (prey) are easier to find and why.

STEP 3

When all hunters have starved, **stand** shoulder-to-shoulder at the end of **Toothpick Territory**. **Walk** slowly forward. **Watch** for "hidden prey." **Make notes** about how many and what color toothpicks are found in this final hunt.

STEP 4

Rejoin your research team and **review** each step in this activity. **Make notes** about what you discovered. Also, **discuss** why the safety rules were important. **Share** and **compare** observations with other research teams.

WHAT HAPPENED?

Notice that the *easier* a color was to see, the easier it was to "catch" that particular toothpick. On the other hand, the colors that were *harder* to see helped the toothpick stay hidden. It's much the same way in the wild. Animals and insects that blend into their surroundings are a lot less likely to become lunch!

Keep in mind that God created many ways for **animals** to protect themselves, and **color** is only one. Some animals (like rabbits) can run and turn very fast. Others (like porcupines) have special **physical defenses** built in. Still others (like skunks) have less subtle techniques to discourage hungry **predators**. All of these things help maintain the important **balance** between predator and **prey**.

WHAT WE LEARNED

 Describe the research site (habitat) your teacher created. How did this affect which toothpicks were easiest to find?

_____

_____

_____

_____

_____

_____

 What color toothpick was found most often? What color toothpick was found least often? Why?

_____

_____

_____

_____

_____

_____

3 Suppose the toothpicks were relocated to a different habitat (like a bright red piece of carpet). How would this affect the results? Why?

_____

_____

_____

_____

_____

_____

4 Characteristics that lead to survival become more common over time. Why? If these toothpicks had offspring, what color might be more common in the next generation?

_____

_____

_____

_____

_____

_____

5 Besides color, what are some other ways animals, birds, and insects protect themselves from predators? Do these protection techniques always work? Why or why not?

_____

_____

_____

_____

_____

_____

CONCLUSION

The relationship between an animal's color and its habitat impacts its potential survival. Characteristics that lead to survival tend to become more common over time.

FOOD FOR THOUGHT

Genesis 1:24-26 God created this world with a complex ecosystem, where the life of one creature interrelates with the lives of many other creatures. For instance, too many predators and the prey all die. Soon the predators starve. But if there are not enough predators, prey will multiply beyond what an area can support. Soon they eat up all the plants, and the prey begin to starve.

This Scripture talks about the creatures God created in his beautiful new world. The last verse reminds us that we are responsible for taking care of this world. Learn to love and care for it. After all, it's the only world we have!

JOURNAL   My Science Notes

LESSON 8

BOSS EYE

FOCUS Eye Dominance

OBJECTIVE To explore brain function through eye dominance

OVERVIEW Are both your eyes equal or is one always the "boss"? In this activity, we'll explore the concept of eye dominance.

WHAT TO DO

STEP 1

Roll a sheet of paper (lengthwise) into a one-inch tube. **Tape** the edge to hold it in place. Keeping both eyes open, **place** the tube over your left eye. **Look** at an object across the room. **Make notes** about how easy it is to see.

STEP 2

Now **place** the tube over your right eye. (Be sure to keep both eyes open.) **Look** at an object across the room. **Make notes** about how easy it is to see.

STEP 3

Switch the tube back and forth until you decide which eye seems easier to use. **Record** the results. (If it was equally easy with either eye, then note that instead.)

STEP 4

Interview at least 10 class members about this activity. **Make notes** about what you discover. Now **share** and **compare** observations with your research team.

WHAT HAPPENED?

Your **eyes** help your **brain** to **perceive** the world around you. Each eye reports to a different half (**hemisphere**) of your brain! The left side of your brain reads **images** from your right eye, and the right side reads images from your left eye. Sounds complicated, but your brain figured out how to combine those images correctly when you were still a baby.

Even though both eyes report what they see, one eye is **dominant** — the eye your brain relies on most. Many left-handed people are left eye dominant, and right-handed folks usually prefer their right eye. But as you probably noticed in your classmates' results, everyone doesn't follow this pattern. In fact, one of the authors of this science series is right-handed, but left-eyed. (That's a little awkward for some things, but a great combination for baseball!)

WHAT WE LEARNED

 Compare Step 1 and Step 2. How were the results similar? How were they different?

_____

_____

_____

_____

_____

_____

 Which eye did you select in Step 3? How difficult was the selection? What factors influenced your decision?

_____

_____

_____

_____

_____

3 Review your notes from Step 4. How many students were right-eye dominant? How many were left-eye dominant? How many indicated no preference?

_____

_____

_____

_____

_____

_____

4 List the right-eye dominant students. Are any of them left-handed? List the left-eye dominant students. Are any of them right-handed?

_____

_____

_____

_____

_____

_____

5 If the right side of a person's brain was injured in a car accident (but their face was okay), which eye might be affected? Why?

_____

_____

_____

_____

_____

_____

CONCLUSION

Even though both eyes report what they see, one eye is dominant. Your right eye works with the left side of your brain, and your left eye works with the right side of your brain.

FOOD FOR THOUGHT

Ephesians 4:14-16 In this activity, you discovered that one of your eyes is dominant. Your body always operates its very best when this eye takes the lead.

This Scripture reminds us that Jesus should always be our leader. Under his direction, the body of the church works together perfectly. Sometimes we forget this, changing our minds about what we believe just because someone has cleverly made a lie sound like the truth. But remember, when Christ is in control of your life, you're always at your very best!

 **JOURNAL**   My Science Notes

NAME _____

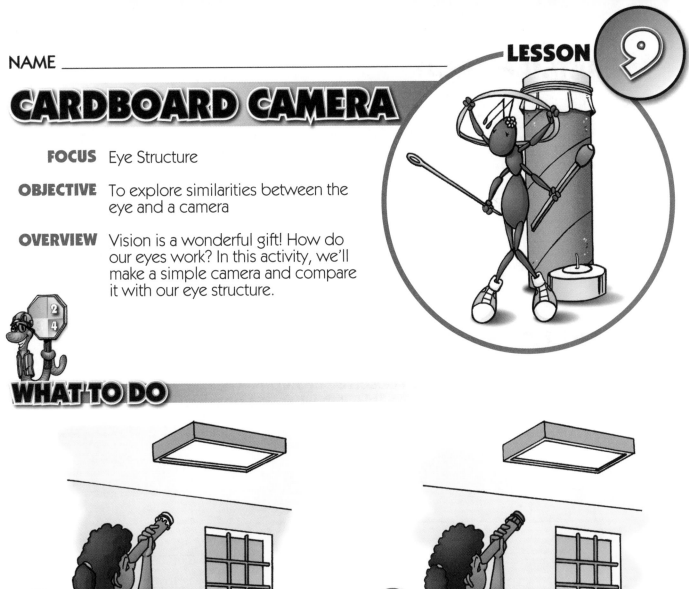

CARDBOARD CAMERA

FOCUS Eye Structure

OBJECTIVE To explore similarities between the eye and a camera

OVERVIEW Vision is a wonderful gift! How do our eyes work? In this activity, we'll make a simple camera and compare it with our eye structure.

WHAT TO DO

STEP 1
Cover the end of a cardboard tube with a single layer of deli paper. (Make sure it's flat and smooth.) **Attach** a rubber band to hold it in place. Now **look** through the tube at a light in the room. **Make notes** about what you see.

STEP 2
Cover the other end of the tube with a piece of construction paper. **Attach** a rubber band to hold it in place. Now **point** the construction paper end toward the light. **Observe** the deli paper end and **make notes** about what you see.

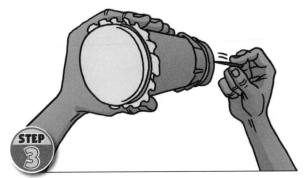

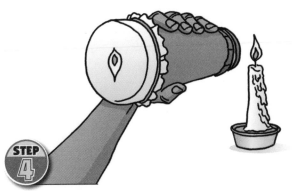

STEP 3
Use a straight pin to **punch** a small hole in the construction paper. **Repeat** Step 2 and **make notes** about what you see. Now **watch** as your teacher lights the candle and darkens the room.

STEP 4
Look through the deli paper and **point** the hole in the construction paper at the candle flame. **Move** the tube back and forth to get a good image. **Make notes** about what you see. **Share** and **compare** observations with your research team.

WHAT HAPPENED?

The purpose of a **lens** (including the one in your **eye**) is to **focus** available **light**. Common applications include telescopes, cameras, binoculars, and microscopes. Some devices contain a complex series of lenses, like a set of funnels and gates. These funnels and gates help control the size and direction of an image.

In a camera, the lens focuses the light on film (or a digital device) which records and stores the **image**. In your eye, the lens focuses light onto the **retina**, a surface used for much the same purpose.

Glasses and contacts simply provide an additional lens that is used to assist the eye's lens when needed.

WHAT WE LEARNED

Describe what you observed in Step 1. What did you see through the deli paper. Compare this with what you observed in Step 2.

_____

_____

_____

_____

_____

_____

_____

Describe what you observed in Step 3. How was this similar to Step 1? How was it different?

_____

_____

_____

_____

_____

_____

_____

Describe what you observed in Step 4. How was this similar to Step 3? How was it different?

_____

_____

_____

_____

_____

_____

Name the two primary parts of the eye modeled in this activity, and describe the purpose of each.

_____

_____

_____

_____

_____

_____

What is the purpose of any lens (including the eye's lens)? Name three common devices that use lenses, and describe their purpose.

_____

_____

_____

_____

_____

_____

_____

CONCLUSION

The purpose of any lens is to focus light. In the eye, the lens focuses light on the retina. The retina then sends the image to the brain for interpretation.

FOOD FOR THOUGHT

John 8:12 In your Cardboard Camera, the deli paper screen was completely dark until light was focused on it. But when the light came into focus, suddenly there was a clear image!

This Scripture reminds us that Jesus is the light of the world. When our eyes are focused on Jesus, everything begins to become clear. Just like the Cardboard Camera, it's easy to lose our focus — and we can end up stumbling through the darkness. Remember to stay focused by spending time with God each day!

JOURNAL   My Science Notes

Forces

In this section, you'll discover how "**push** and **pull**" form the basis for all physical movement. You'll explore simple machines (levers, pulleys). You'll work with Newton's laws of motion. You'll even learn to understand concepts like torque, inertia, and buoyancy.

POUR PROBABILITY

FOCUS Half Life

OBJECTIVE To explore how radioactive elements break down

OVERVIEW Atoms are the same kind of element forever — except radioactive atoms! In this activity, we'll explore probability and simulate a radioactive break down.

WHAT TO DO

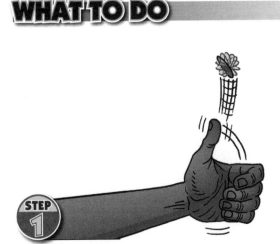

STEP 1

Flip a coin. **Catch** it in one hand, then **turn** it over onto the back of your other hand. Is it heads or tails? **Repeat** this 100 times. **Record** the results of each flip on your journal page. This process simulates something scientists call probability.

STEP 2

Examine the paper your teacher gives you. **Make notes** about what you see. Now use a paper punch to **create** exactly 100 tiny black and white disks. **Place** all the disks in a small container and **shake** it thoroughly.

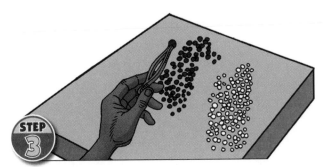

STEP 3

Pour the disks onto your work surface and **spread** them out. **Place** all the "white side up" disks in a pile. (Tweezers make them easier to grab.) Once sorted, **record** the number of each. This "sampling" represents a radioactive element's half-life.

STEP 4

Place the dark disks back in the container and **shake** them up again. **Repeat** step #3 and **record** the results. **Repeat** Steps 3 and 4 until all the dark disks are gone. Now **review** each step in this activity. **Share** and **compare** observations with your research team.

❓ WHAT HAPPENED?

The inside of an atom (nucleus) contains **protons** and **neutrons**. Normally the **nucleus** is incredibly stable, so an **atom** stays the same **element** forever. But some kinds of atoms are **unstable**, constantly breaking down into more **stable** forms, releasing potentially dangerous **energy** in the process. These are **radioactive** atoms.

In this activity, we simulated that breakdown. Each time a white disk fell face up, it represented the **probability** (chance) that one atom of an element would break down. The time it takes for half the atoms of an element to break down is called its **half-life**. In our **simulation** this was probably less than a minute. In the real world, half-lives of elements vary enormously! Some half-lives are measured in tiny fractions of a second, others are measured in eons of time.

❓ WHAT WE LEARNED

 Compare your coin toss probability from Step 1 with your disk probabilities in the other steps. How were they similar? How were they different?

_____

_____

_____

_____

_____

_____

 Review your results from each shake and toss. How many white disks and how many black disks occurred in each? What does this tell you about probability?

_____

_____

_____

_____

_____

_____

3 Describe the condition of an atom's nucleus in a radioactive element. Why could this make it dangerous?

_____

_____

_____

_____

_____

_____

_____

4 Define a simulation. How can simulations help us understand the real world?

_____

_____

_____

_____

_____

_____

_____

5 Define the half-life of an element. How much variation is there in the half-life of different elements?

_____

_____

_____

_____

_____

_____

_____

CONCLUSION

Although most atoms are incredibly stable, radioactive atoms are constantly breaking down, releasing potentially dangerous energy. The time it takes for half an element's atoms to break down is known as the half-life.

FOOD FOR THOUGHT

James 5:16 One radioactive atom breaking down releases some energy. When large numbers of radioactive atoms break down, it releases an enormous amount of force! A nuclear power plant can produce enough electricity for an entire city, and provides a good example of the collective force of tiny, individual changes.

This Scripture tells us that the earnest prayer of just one believer has great power. Imagine what might happen if we all begin to pray together! Prayer links us with the most powerful force in the universe — our creator God. Why not experience the power of prayer today?

JOURNAL   My Science Notes

WELDED WATER

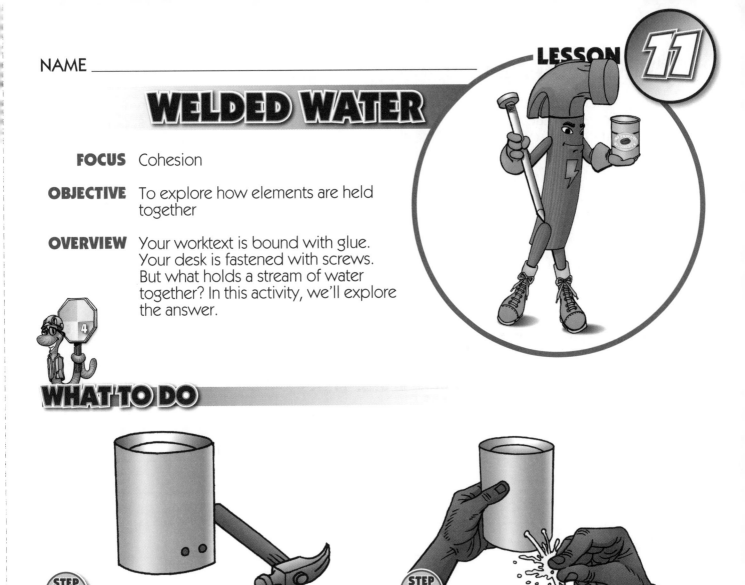

FOCUS Cohesion

OBJECTIVE To explore how elements are held together

OVERVIEW Your worktext is bound with glue. Your desk is fastened with screws. But what holds a stream of water together? In this activity, we'll explore the answer.

WHAT TO DO

STEP 1

Using a hammer and nail, **punch** two holes just above the bottom of a can (see illustration). The holes must be the same size and exactly 1/4 inch apart. **Examine** the can and **make notes** in your journal. **Predict** what will happen if you fill the can with water.

STEP 2

Seal the holes with your fingers and **fill** the can with water. Hold it in front of you, holes out. **Remove** your fingers and **observe** the two streams. Now quickly "**pinch**" the streams together. (This may take practice.) **Make notes** about what happens.

STEP 3

Repeat Step 2, pinching the two streams into one. Now quickly **wipe** one finger down across the holes to briefly interrupt the flow of water. **Make notes** about what happens to the water.

STEP 4

Make sure everyone on your team has a turn working with the streams of water. Now **review** each step in this activity. **Make notes** about what you observed. **Share** and **compare** observations with your research team.

WHAT HAPPENED?

When **molecules** are held together by **mutual attraction**, it's known as **cohesion**. Cohesion causes the molecules to act like tiny magnets, attracting and sticking to each other. When you "welded" the streams of water, you actually just pinched them close enough for cohesion to turn them into one stream. When you wiped your finger across the holes, you broke the cohesion. This resulted in separate streams.

Cohesion is also what allows some insects (like pond skaters and water striders) to walk on water. The water molecules at the surface are attracted to each other and also to the molecules just below them. This forms a kind of "skin" at the surface, a type of cohesion that is known as **surface tension**.

WHAT WE LEARNED

What did you predict in Step 1? How did this prediction reflect what actually happened?

_____

_____

_____

_____

_____

_____

_____

Describe what happened to the two streams in Step 2. Why did this happen? What is the term used to describe this kind of force?

_____

_____

_____

_____

_____

_____

_____

3 Describe what happened to the streams in Step 3.
What caused this to occur?

_____

_____

_____

_____

_____

_____

4 What might have happened if we'd used a liquid with no
cohesion? Why?

_____

_____

_____

_____

_____

_____

5 Name another example of cohesion and describe how it works.

_____

_____

_____

_____

_____

_____

CONCLUSION

Cohesion is a force that holds elements together through mutual attraction. Cohesion at the top of a body of water is called surface tension.

FOOD FOR THOUGHT

Acts 1:12-14 When the streams of water were rushing from the can, they were each going their own way. Each stream was following its own path, and they definitely weren't working together. But when the streams made contact with each other, their combined power made a much greater impact!

Prayer is a very important part of spiritual health. Having a personal prayer life is very important, but don't overlook the power of praying with others. This Scripture tells us how the early church members joined together in prayer. If one prayer is powerful, imagine the strength of many prayers!

JOURNAL   My Science Notes

_____

_____

_____

_____

_____

_____

_____

_____

_____

_____

_____

_____

_____

_____

_____

_____

_____

_____

_____

LESSON 12

BALLOON BUST!

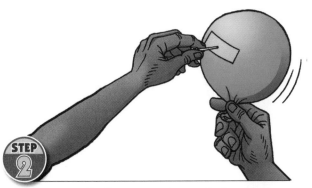

FOCUS Stress

OBJECTIVE To explore how stress affects materials

OVERVIEW Sharp pins pop balloons! Everyone knows that — or do they? In this activity, we'll explore how things aren't always as simple as they appear.

WHAT TO DO

STEP 1

Inflate a balloon so it's full, but not too tight. **Observe** the balloon closely. **Predict** what will happen if you touch the balloon with a sharp pin. Now **stick** the balloon in the side with the pin. **Record** the results on your journal page.

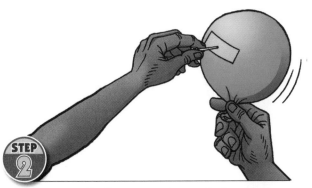

STEP 2

Inflate a second balloon. **Attach** a two inch strip of tape to the side of the balloon. (Make sure the tape is completely flat.) **Predict** what will happen if you push the pin through the tape. Now slowly **push** the pin through the tape. **Record** the results.

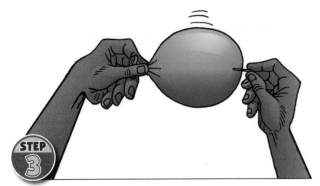

STEP 3

Inflate a third balloon. **Locate** the exact bottom of the balloon (directly opposite the neck). **Make notes** about how this area looks different from the side of the balloon. Now slowly **push** the pin into this area. **Record** the results.

STEP 4

Review each step in this activity. How were the steps similar? How were they different? **Share** and **compare** observations with your research team.

WHAT HAPPENED?

An inflated balloon normally pops when a pin hits it. This is because its tightly stretched latex surface is under a lot of **stress**! (In this case, the stress comes from the **force** of the air you **pushed** into the balloon.) When a pin makes a hole, the stressed latex can't deal with the rapidly rushing air, so it tears and the balloon explodes!

Some **materials** can handle a lot of stress. Others can be modified to increase their ability to handle stress. Sometimes stress areas can be bypassed altogether.

In Step 2, you increased an area's ability to handle stress. By adding the adhesive tape, you gave the latex some backing and support. In Step 3, your bypassed the stressed areas, focusing on a spot where the latex was not as stressed (stretched). When the pin made a hole, the material didn't tear, and the balloon didn't pop!

WHAT WE LEARNED

 What did you predict in Step 1? How did this prediction reflect what actually happened?

_____

_____

_____

_____

_____

_____

_____

 What did you predict in Step 2? How did this prediction reflect what actually happened?

_____

_____

_____

_____

_____

_____

_____

3 In Step 3, how did the surface of the balloon vary? What happened when you pushed the pin into the bottom? Why?

_____

_____

_____

_____

_____

_____

4 Explain why every team's balloon exploded in Step 1, but most balloons didn't explode in Step 2 or Step 3.

_____

_____

_____

_____

_____

_____

5 Describe how adding a piece of adhesive tape changed the balloon's ability to handle stress.

_____

_____

_____

_____

_____

_____

CONCLUSION

Stretching a material creates stress. Some materials handle stress well. Others can be modified to increase their stress-handling ability.

FOOD FOR THOUGHT

Daniel 6:3-24 Before this activity, everyone in your class probably thought that pins always pop balloons. But your teacher knew that wasn't true. The end result didn't fool your teacher, but it probably was a big surprise for you!

Daniel's enemies (and even his friends) were certain he would be eaten by those hungry lions. But God knew that wasn't true. God had other plans for Daniel. In the end, it was Daniel's enemies who became lunch for the lions! Which is more trustworthy — what everyone thinks is true, or what God says is true?

JOURNAL My Science Notes

WEIGHTLESS WONDER

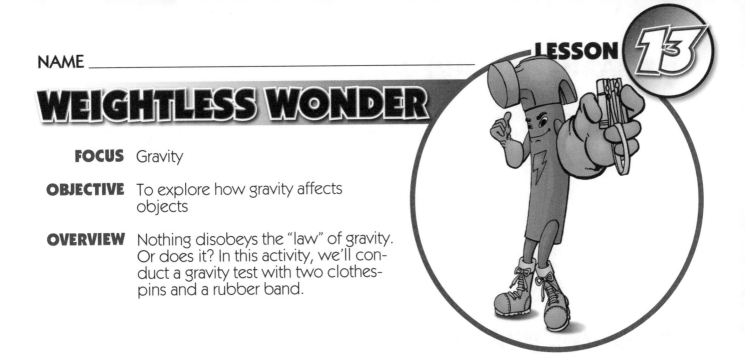

FOCUS Gravity

OBJECTIVE To explore how gravity affects objects

OVERVIEW Nothing disobeys the "law" of gravity. Or does it? In this activity, we'll conduct a gravity test with two clothespins and a rubber band.

WHAT TO DO

STEP 1

Pick up a clothespin. **Hold** your arm straight out in front of you. **Release** the clothespin. **Observe** what happens, and **make notes** in your journal.

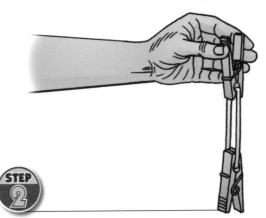

STEP 2

Clip two clothespins onto a rubber band (see illustration). **Hold** one clothespin, letting the other hang by the rubber band. **Observe** the rubber band. **Draw** a picture of its shape in your journal.

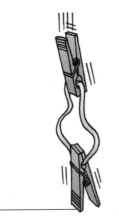

STEP 3

Predict what will happen to the shape of the rubber band if you let go of the top clothespin. **Ask** a team member to watch closely from the side as you drop the top clothespin. **Discuss** what happened to the shape of the rubber band. **Make notes** in your journal.

STEP 4

Review each step in this activity. **Discuss** why the bottom clothespin appeared to be momentarily weightless. **Make notes** in your journal. **Share** and **compare** observations with your research team.

WHAT HAPPENED?

When you dropped the top clothespin, it began to fall immediately. But the bottom clothespin hesitated — apparently weightless for a moment! What really happened is that **inertia** (and the **pull** of the rubber band) came into play. **Gravity** pulled on the clothespins equally. But since you let go of the top clothespin first, gravity pulled it down first while other forces kept the bottom clothespin in place a moment longer.

Let's look at this again. Before you released the clothespin, your upward pull was exactly equal to gravity's downward pull. That's why the clothespins didn't move. But when you let go of the top clothespin, the upward pulling force was gone! The result was a chain reaction. The top clothespin fell, the rubber band returned to its original shape, this pulled on the bottom clothespin, and the combination of pull and inertia made the bottom clothespin hesitate before falling.

WHAT WE LEARNED

 Describe what happened to the clothespin in Step 1. What force affected it?

 Why didn't the clothespins fall in Step 2? What were the two opposing forces involved?

What did you predict in Step 3? How did this prediction reflect what actually happened?

_____

_____

_____

_____

_____

_____

In Step 3, what force caused the rubber band to stretch? What two things caused it to bulge out?

_____

_____

_____

_____

_____

_____

Describe how various forces made the bottom clothespin appear momentarily weightless.

_____

_____

_____

_____

_____

CONCLUSION

Forces can create the illusion of momentary weightlessness, but gravity and inertia still affect everything on Earth.

FOOD FOR THOUGHT

Job 26:7-14 Everything must obey the law of gravity. In this activity, gravity was exerting a constant pull on every object. No matter where you go in the Universe, gravity is present to some degree, constantly pulling you.

This Scripture reminds us that God is present everywhere. Like gravity, God is constantly pulling everyone closer, seeking to draw them into a relationship with him. Remember, no matter what happens, no matter how frightened or sad or lonely you become, God is always there for you.

 **My Science Notes**

_____

_____

_____

_____

_____

_____

_____

_____

_____

_____

_____

_____

_____

_____

_____

_____

_____

_____

_____

_____

NAME _____

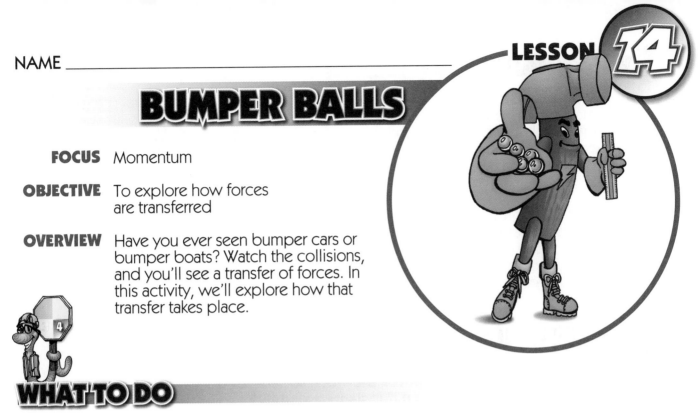

BUMPER BALLS

FOCUS Momentum

OBJECTIVE To explore how forces are transferred

OVERVIEW Have you ever seen bumper cars or bumper boats? Watch the collisions, and you'll see a transfer of forces. In this activity, we'll explore how that transfer takes place.

WHAT TO DO

STEP 1

Check your work surface for "level" by placing a metal ball in the center. If the ball rolls in any direction, **ask** your teacher for help. Now **place** a ruler on the surface. **Place** a metal ball on the ruler at the 4 inch mark, and another at the 1/2 inch mark.

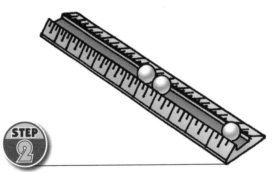

STEP 2

Gently **flick** Ball #1 (on the end) toward Ball #2 (at the 4 inch mark). **Record** the results. **Replace** the balls, then **add** a third ball (see illustration). It must touch Ball #2! Now gently **flick** Ball #1 toward Ball #2 as before. **Record** the results.

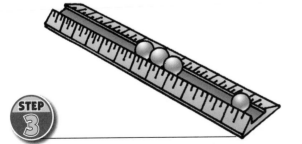

STEP 3

Replace all three balls, then **add** a fourth ball (see illustration). Balls #2, #3, and #4 must touch! Now gently **flick** Ball #1 toward Ball #2 as before. **Record** the results.

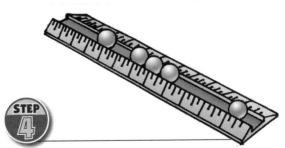

STEP 4

Repeat Step 3, but move Ball #4 about 1/4 inch away from Ball #3 so they no longer touch. **Record** the results. Now **review** each step in this activity. **Make notes** about what you observed. **Share** and **compare** observations with your research team.

FORCES **65**

WHAT HAPPENED?

Did you find the results surprising? When the moving ball hit the target (transferring its momentum), only the last ball in line moved! No matter how many balls were in line, only the last one took off. That's because each stationary ball simply transferred the original momentum to the next one in line. Since the last ball had nothing to hold it back, it rolled away. Notice that the balls had to be touching, or the **force** couldn't transfer.

Any moving object has **momentum**. Momentum is a combination of how big an object is (**size**) and how fast it's moving (**speed**). If two objects are identical, then the faster one has more momentum. But a small object moving extremely fast (like a bullet) could have much more momentum than a large object moving slowly (like a sumo wrestler).

WHAT WE LEARNED

 Why is it important for the work surface to be level for this activity? What might happen if it slanted slightly?

_____

_____

_____

_____

_____

_____

_____

 What happened to Ball #2 in the first part of Step 2? What happened in the second part of Step 2? Why were the results different?

_____

_____

_____

_____

_____

_____

_____

Describe what happened in Step 3. Why didn't Ball #2 and Ball #3 move much? What happened to Ball #4? Why?

_____

_____

_____

_____

_____

_____

4 Why was it important for the balls to be touching in Step 2, Step 3, and Step 4? What would have happened to the momentum if there were gaps between the balls?

_____

_____

_____

_____

_____

_____

_____

5 Momentum is a combination of what two factors? How can a small object have more momentum than a large one?

_____

_____

_____

_____

_____

_____

CONCLUSION

Momentum can be transferred between objects. Momentum is a combination of an object's size and its speed.

FOOD FOR THOUGHT

Isaiah 30:18, 21 Once you understood the principles involved, you were able to control the direction of momentum. By setting up the balls just so, you were able to transfer forces effectively and keep the balls from falling off the work surface.

This Scripture reminds us that God is waiting for us to let him take control of our lives. God understands and controls the forces around us. When we trust ourselves to his care, he is able to guide us to the correct path and keep us from falling. Our part is to learn to listen to God and follow his instructions.

 **JOURNAL**   **My Science Notes**

NAME _____

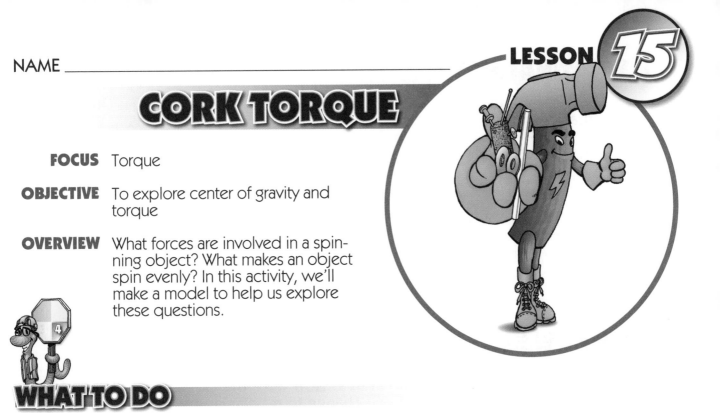

CORK TORQUE

FOCUS Torque

OBJECTIVE To explore center of gravity and torque

OVERVIEW What forces are involved in a spinning object? What makes an object spin evenly? In this activity, we'll make a model to help us explore these questions.

WHAT TO DO

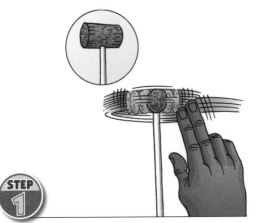

STEP 1

Stick a pin in the side of the cork near the center. **Hold** a straw vertically and **slip** the head of the pin into the straw (see illustration). **Spin** the cork by gently tapping one end. If it falls, **relocate** the pin and try again. **Record** the results.

STEP 2

Attach a small washer to one end of the cork with a push pin. Try to **spin** the cork by gently tapping one end. **Record** the results. **Relocate** the pin and try again until the cork spins freely. **Make notes** about the changes you made.

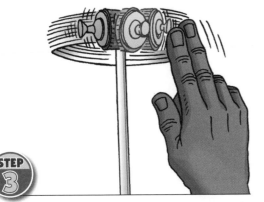

STEP 3

Replace the small washer with the bigger washer. Try to **spin** the cork again by gently tapping one end. **Record** the results. **Relocate** the pin and try again until the cork spins freely. **Make notes** about the changes you made.

STEP 4

Review all the steps in this activity. **Make notes** about the challenge of spinning the cork each time and the adjustments you made to correct any problems. **Share** and **compare** observations with your research team.

FORCES **69**

WHAT HAPPENED?

A twisting **force** is called **torque**. To **balance** torque, you have to find the **center of gravity**. Putting a pin in the exact center of the cork helped it spin freely. But when you added weight to one side, the center of gravity changed. You had to move the pin to get the cork to spin freely again.

An object's center of gravity is related to its **shape** and **mass**. A ruler balances easily at the center because it's almost identical on both sides of that point. But a baseball bat is much heavier on one end. You have to move toward the heavy end of the bat to find its center of gravity. Balancing a teeter-totter is similar, except instead of moving the balance point, we move the **weight** toward or away from the middle to shift the center of gravity.

WHAT WE LEARNED

 Compare Step 1 with Step 2. How were they similar? How were they different? What did you change to rebalance the cork?

_____

_____

_____

_____

_____

_____

_____

 Describe Step 3. What did you do with the support pin to make the cork spin freely? Why?

_____

_____

_____

_____

_____

_____

_____

3 What is a "twisting force" called? What might happen if this force is not balanced?

_____

_____

_____

_____

_____

_____

4 If you added an even bigger washer in Step 3, which way would the support pin need to move? Why?

_____

_____

_____

_____

_____

_____

5 Using what you've learned in this activity, explain why the spinning parts of an engine must be balanced.

_____

_____

_____

_____

_____

_____

CONCLUSION

Twisting force is called torque. Balancing torque requires locating the center of gravity — the point around which a spinning object turns freely.

FOOD FOR THOUGHT

Proverbs 12:25 Once you understand how torque works, balancing the cork isn't too hard. But the pin only has to be pushed a little bit off center to start creating problems again!

Many people face tragedies (a death, an accident, etc.) that throw them off balance. But this Scripture reminds us that a word of encouragement does wonders! As God's children, we must always be ready to share his love and caring compassion with others — especially when their hearts are heavy. Kind words don't make a tragedy go away, but they can often help someone recover that needed balance.

JOURNAL My Science Notes

_____

_____

_____

_____

_____

_____

_____

_____

_____

_____

_____

_____

_____

_____

_____

_____

_____

_____

_____

_____

BUOYANT BOAT

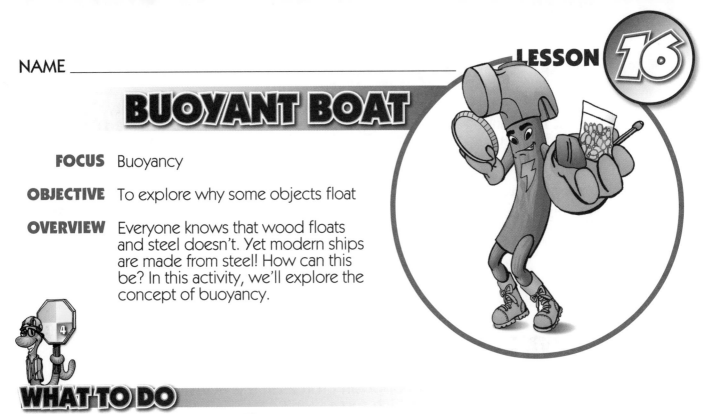

FOCUS Buoyancy

OBJECTIVE To explore why some objects float

OVERVIEW Everyone knows that wood floats and steel doesn't. Yet modern ships are made from steel! How can this be? In this activity, we'll explore the concept of buoyancy.

WHAT TO DO

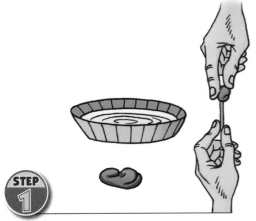

STEP 1

Fill a pie pan with water. **Hold** a chunk of clay at the water's surface, then **release** it. **Record** the results. **Pinch** off a piece of clay about the size of a pea. **Push** it onto the end of a matchstick to make a "matchstick sailor."

STEP 2

Design a boat using the remaining clay. **Set** your boat on the surface of the water and **release** it. **Record** the results. If your boat didn't float, redesign it. Now carefully **stick** your "sailor" into the clay. **Adjust** as necessary to keep your boat stable.

STEP 3

Observe your boat for a few minutes. **Record** your observations. Now slowly **fill** your boat with unpopped popcorn — one piece at a time. **Record** how many kernels your team's boat holds before sinking.

STEP 4

Review each step in this activity. **Make notes** about what you've observed. **Share** and **compare** observations with your research team.

WHAT HAPPENED?

All **matter** on Earth is constantly being **pulled** downward by **gravity**. To keep an object from sinking in the water, gravity has to be overcome (**pushed**) by another **force**. The opposing force we see pushing in this activity is called **buoyancy**.

When you dropped the chunk of clay in Step 1, it sank. When you molded the same clay into a boat shape, it began to float! How is this possible? The boat shape allowed the clay to **displace** (shove aside) the water. If the water displaced **weighs** more than boat and its contents, the boat floats. But if the boat and contents weigh more than the water (as it did when you added enough corn), then the boat sinks.

WHAT WE LEARNED

 What were the two opposing forces in this activity?
What determined which force won?

_____

_____

_____

_____

_____

_____

_____

 Compare the clay used in Step 1 and Step 2. How was it the same?
How was it different?

_____

_____

_____

_____

_____

_____

_____

3 How much corn did your boat hold before sinking? How were the more successful boats different from the others?

_____

_____

_____

_____

_____

_____

_____

4 When a boat builder talks about the boat's "displacement," what does this mean?

_____

_____

_____

_____

_____

_____

_____

5 Based on what you've learned, why would a boat sink if a hole let water in?

_____

_____

_____

_____

_____

_____

_____

CONCLUSION

The displacement of water can create buoyancy. Buoyancy and gravity are opposing forces.

FOOD FOR THOUGHT

Acts 27:39-41 Your clay boat was hardly seaworthy, but it did manage to float in your pie pan pond. As long as the force of buoyancy kept holding it up, your popcorn cargo survived. When the boat was overloaded, water poured in and it sank.

Scripture tells of a frightening shipwreck Paul experienced. As the ship struck the rocks, water began to rush in. The ship no longer had the support of buoyancy and it sank. People can be like that, too. Without God's support, the burdens of life can overload them as trials and obstacles rush in and drag them down. Only God's love can help us weather the storms of life!

JOURNAL My Science Notes

FINGERCUFF GOOD

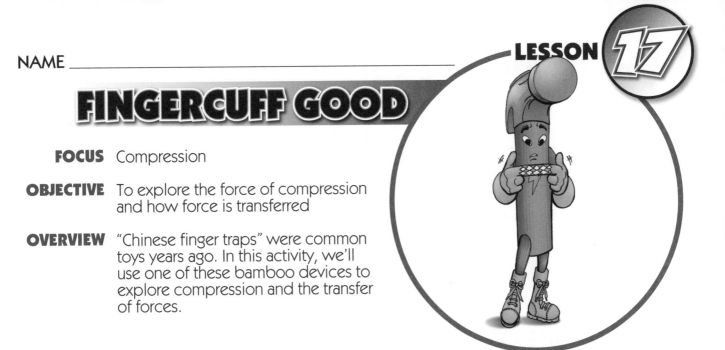

FOCUS Compression

OBJECTIVE To explore the force of compression and how force is transferred

OVERVIEW "Chinese finger traps" were common toys years ago. In this activity, we'll use one of these bamboo devices to explore compression and the transfer of forces.

WHAT TO DO

STEP 1

Examine the finger trap carefully. **Make notes** in your journal about what you see. Now **hold** a piece of cloth by its opposite edges. Gently **twist** the cloth while pulling. **Compare** the way the fibers move to the action of the finger trap.

STEP 2

Push your right index finger firmly into one end of the finger trap, then **push** your left index finger into the other end. Now try to **pull** both fingers out of the finger trap. (Don't pull too hard!) **Observe** what happens.

STEP 3

Let's try again. This **time** push your fingers toward each other, **hold** the finger trap loosely in place with your thumbs, then gently **slide** your fingers out. **Make notes** about why you think this method was more successful.

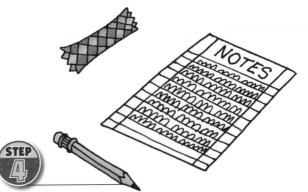

STEP 4

Review each step in this activity. **Make notes** about how the finger trap's shape changed in Step 2 and Step 3. **Share** and **compare** observations with your research team.

WHAT HAPPENED?

When you **pushed** your fingers into the trap, it changed shape slightly to let them in. But when you started to **pull** them out again, the trap changed shape to grab your fingers! As you pulled harder, the **force** of your **muscles** was **transferred** to the trap, causing it to **twist** and compress even more tightly around your fingers.

The trap works because of the interlocking, woven strips. Pushing in separates the strips slightly, but pulling out binds them more tightly together. As you discovered, the only way to get out was to understand the forces involved and work with them, not against them!

WHAT WE LEARNED

 Compare the action of the finger trap with the twisting cloth in Step 1. How were they similar? How were they different?

_____

_____

_____

_____

_____

_____

 What happened when you tried to pull your fingers out of the trap in Step 2? Why?

_____

_____

_____

_____

_____

_____

What happened when you tried to pull your fingers out of the trap in Step 3? Why?

_____

_____

_____

_____

_____

_____

How did the shape of the finger trap change in each step? What caused this to occur?

_____

_____

_____

_____

_____

_____

_____

Describe how this activity demonstrated a transfer of force. Where did the force originate?

_____

_____

_____

_____

_____

_____

CONCLUSION

Forces can be transferred from one place to another.

FOOD FOR THOUGHT

Galatians 3:21-22 With a finger trap, the harder you try to force your way out, the more you're trapped! You can only be set free by understanding the forces involved and following the rules.

Life is like a finger trap. The harder we try without God's help, the more difficult things become. We're quickly caught in the awful grip of sin! But by spending time with God, we learn to understand the forces involved. Give in to the finger trap, and your fingers are free. Give in to God, and your soul is free!

JOURNAL   My Science Notes

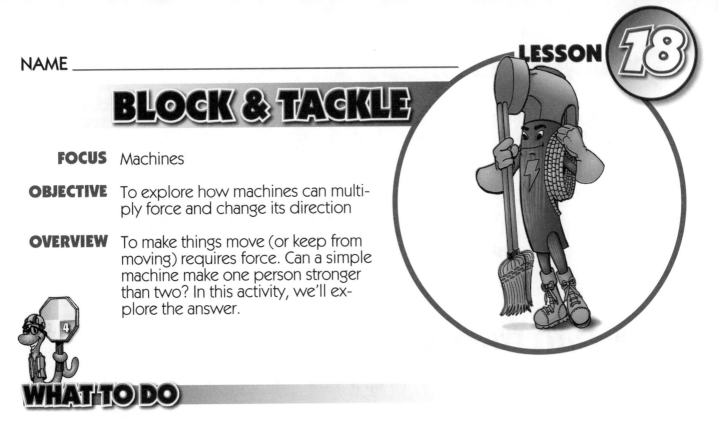

BLOCK & TACKLE

FOCUS Machines

OBJECTIVE To explore how machines can multiply force and change its direction

OVERVIEW To make things move (or keep from moving) requires force. Can a simple machine make one person stronger than two? In this activity, we'll explore the answer.

WHAT TO DO

STEP 1

Hold two broomsticks in front of you, one in each hand. **Ask** one team member to grab the right broom, and another the left. When you say "Go!" have them gently **pull** the broomsticks apart. **Make notes** about what happens.

STEP 2

Ask two team members to **hold** the sticks about two feet apart. **Tie** one end of the rope to a stick's top, **wrap** it around the second stick, then back to the first (see illustration). **Hold** the rope tight and **ask** your team members to try to **pull** the sticks apart. **Record** the results.

STEP 3

Repeat Step 2, but this time **wrap** the rope back and forth several times (see illustration). **Ask** your team members to try to **pull** the sticks apart again. **Record** the results.

STEP 4

Repeat Step 3, but this time have your team members try to hold the sticks steady while you **tighten** the rope. **Take up** as much rope as you can. Now **review** each step and **make notes** about differences. **Share** and **compare** observations with your research team.

WHAT HAPPENED?

The "block and tackle" you created is actually a **compound machine** made from two **simple machines** — two different **wheels** and **axles**. In this activity, the broomsticks played the role of two axles and the rope acted like a set of wheels. Machines with block and tackle arrangements can lift or hold tremendous amounts of **weight** with minimal amounts of **force**.

Notice that in addition to increasing the force, the machine changed the **direction** of the force. (The two people on the broomsticks were **pulling** in different directions from the person controlling the rope.) A car jack is a similar example. A relatively light downward **push** will raise up a very heavy car. The force provided by a person's muscles is both **multiplied** *and* sent in a different direction.

WHAT WE LEARNED

 Compare Step 2 and Step 3. How were they similar? How were they different?

_____

_____

_____

_____

_____

_____

_____

 Compare Step 3 and Step 4. How were they similar? How were they different?

_____

_____

_____

_____

_____

_____

_____

3 Describe what happened in Step 4.
How did this demonstrate the transfer of forces?

_____

_____

_____

_____

_____

_____

4 What two simple machines did the rope/broomstick arrangement imitate? How did it change the direction of a force?

_____

_____

_____

_____

_____

_____

5 Give another example of a machine that either multiplies force or changes its direction. Tell how it works.

_____

_____

_____

_____

_____

_____

CONCLUSION

Machines can multiply force. They can also change the direction of a force.

FOOD FOR THOUGHT

Isaiah 40:29-31 In this activity, we saw how a simple machine was able to multiply the force of one person so they could overpower the strength of two. Machines are wonderful devices to help us increase our force.

Of course, no machine can equal the amazing power that comes from God! This Scripture reminds us that we all have times when we're tired or weak, ready to give up. But if you spend time every day learning to trust him, God can fill you with life-giving strength.

JOURNAL   My Science Notes

_____

_____

_____

_____

_____

_____

_____

_____

_____

_____

_____

_____

_____

_____

_____

_____

_____

_____

_____

Earth Science

Earth Science is the study of **earth** and **sky**. In this section, you'll explore the structure of our planet (rocks, crystals, volcanos), the atmosphere (air, clouds), and related systems (water cycles, air pressure, weather). Grades 7 and 8 reach out even further with a look at the solar system and stars.

NAME _____

 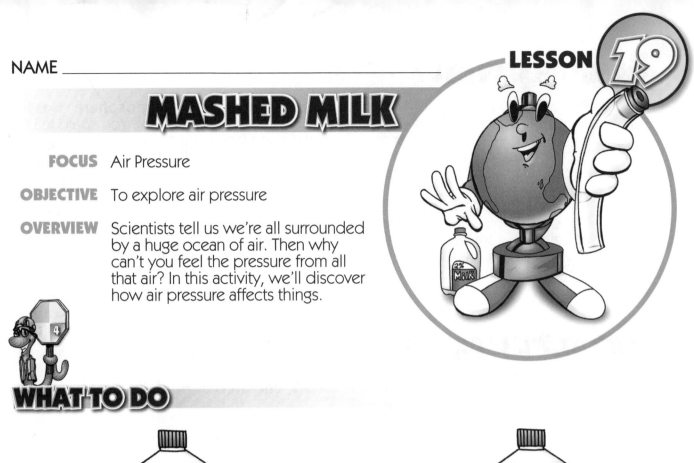

MASHED MILK

FOCUS Air Pressure

OBJECTIVE To explore air pressure

OVERVIEW Scientists tell us we're all surrounded by a huge ocean of air. Then why can't you feel the pressure from all that air? In this activity, we'll discover how air pressure affects things.

WHAT TO DO

STEP 1

Place an empty plastic milk jug on your work surface. **Observe** the jug's shape. **Make notes** in your journal about what you see.

STEP 2

Slowly **fill** the jug with water. Make sure that it's completely full and that no room is left for air. **Observe** the jug's shape. **Make notes** about what you see.

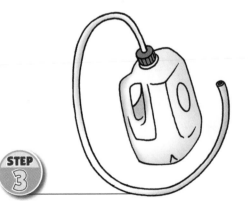

STEP 3

Push the hose into the one-hole stopper. Now **insert** the stopper into the jug's opening. (The hose must be tight in the stopper, and the stopper tight in the jug.) **Observe** the jug's shape. **Make notes** about what you see.

STEP 4

Flip the jug upside down over the sink, letting the water drain out. (If the water is reluctant, gently squeeze the jug — but don't squeeze too hard or the stopper will pop out!) **Observe** the jug's shape. **Share** and **compare** observations with your research team.

EARTH **87**

WHAT HAPPENED?

The empty jug didn't crush because the **pressure** of the air (**atmosphere**) was the same inside and out. Scientists call this kind of balance **equilibrium**. You could tell the air pressure was equal because the jug didn't collapse. Then you filled the jug with water. Although the air was completely removed, the jug still didn't collapse since the air pressure outside was the same as the water pressure inside.

Then you flipped the jug upside down, and the plastic tubing let the water escape. This caused the water pressure inside to begin dropping. No air could get back in to **equalize** the pressure! As the pressure **pushing** out became weaker and weaker, it allowed the pressure outside to push in and crush the container.

WHAT WE LEARNED

 Describe the jug in Step 1, Step 2, and Step 3. How did it look similar? How did it look different?

 Why didn't air pressure crush the jug in Step 1? Why didn't it crush the jug in Step 2?

3 Explain why the jug crushed in Step 4.

_____

_____

_____

_____

_____

_____

_____

4 If you opened a large hole in the bottom of the upside-down jug while it was emptying, would it still have crushed? Why or why not?

_____

_____

_____

_____

_____

_____

_____

5 Based on what you've learned, why do most gasoline cans have a small hole opposite the spout?

_____

_____

_____

_____

_____

_____

 CONCLUSION

Balanced pressures create equilibrium. Major differences in pressure can create strong forces.

 FOOD FOR THOUGHT

Psalm 29:11 Your milk jug is a good example of what happens when pressures push in. When the inside becomes weak and the pressure from the outside is strong, the jug collapses.

People are a lot like that milk jug. We live in a world full of problems and pressures. Unless we learn to rely on God to make us strong inside, those outside pressures can crush us. Spend time getting to know God better, and he will fill you with the strength to withstand the world.

 JOURNAL   My Science Notes

MAXIMUM MARSHMALLOW

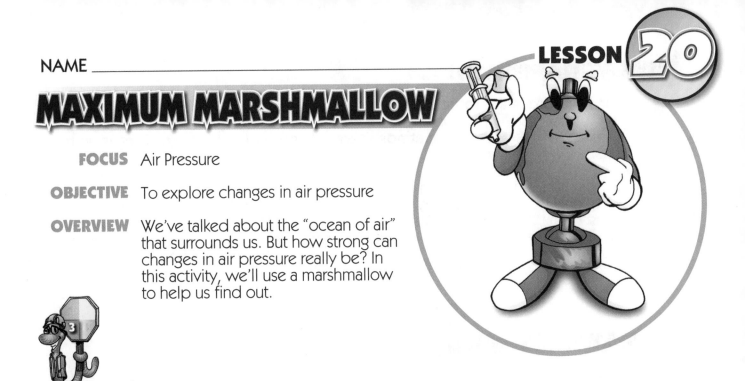

FOCUS Air Pressure

OBJECTIVE To explore changes in air pressure

OVERVIEW We've talked about the "ocean of air" that surrounds us. But how strong can changes in air pressure really be? In this activity, we'll use a marshmallow to help us find out.

WHAT TO DO

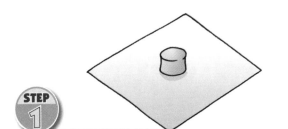

STEP 1

Place a marshmallow on your work surface. **Observe** it closely, looking for any evidence that this marshmallow might suddenly change size! **Make notes** in your journal about what you see.

STEP 2

Pick up the syringe and **pull** the handle all the way out of the tube. Carefully **place** the marshmallow in the tube. **Replace** the handle and slowly **push** down until it almost touches the marshmallow. **Observe** the marshmallow. **Make notes** about what you see.

STEP 3

Continue observing the marshmallow as you **pull** the handle almost out, then slowly **push** it down again until it almost touches the marshmallow. **Repeat** two or three times. **Make notes** about what you see.

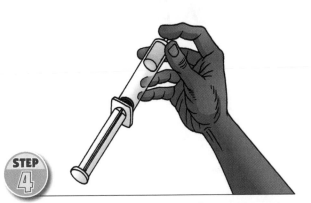

STEP 4

Plug the tip of the syringe with your finger. (The seal must be tight for this to work!) Slowly **pull** the handle back. **Observe** the marshmallow. Now **push** the handle back in. **Observe** the marshmallow. **Repeat** until everyone has had a turn. **Share** and **compare** observations with your research team.

WHAT HAPPENED?

Air pressure is the **force** of the air constantly **pushing** all around us. Except during violent weather, we usually don't have an opportunity to see big changes in air pressure. That's because the **atmosphere** is huge, and air pressure changes normally happen very slowly.

By using the syringe, we shrank the "atmosphere" to the size of the tube. This allowed us to change air pressure very quickly. Since a marshmallow is mostly air (that's why it's so soft), it's a good indicator of big changes in air pressure. When we **pulled** the syringe handle out, there was less air pressure in the tube, so the marshmallow **expanded** (because of the air trapped inside it). When we **pushed** the handle in, air pressure increased, so the marshmallow (and the air inside it) was **compressed** into a smaller size.

WHAT WE LEARNED

1 Compare the marshmallow in Step 2 and Step 4. How was it similar? How was it different?

_____

_____

_____

_____

_____

_____

2 Why was it important to push the handle close to the marshmallow in Step 2? What did this allow us to better control?

_____

_____

_____

_____

_____

_____

3 What is the name of the "ocean of air" that surrounds us? Why don't we see air pressure changes in it?

_____

_____

_____

_____

_____

_____

4 How did the syringe help us see changes in air pressure? How did the marshmallow help?

_____

_____

_____

_____

_____

_____

5 Would a peanut have worked just as well for this activity? Why or why not?

_____

_____

_____

_____

_____

_____

CONCLUSION

Changes in air pressure can create movement. Huge changes in air pressure can have a tremendous effect our weather.

FOOD FOR THOUGHT

Judges 16:28-30 In this activity, your pushing and pulling definitely had an effect on the marshmallow. Although you couldn't see the force itself, the effects of the force were clearly visible from the changes that occurred.

Scripture tells us Samson turned away from God. He was deceived, weakened, imprisoned, and even blinded, But in his weakest hour, Samson remembered the invisible force of God that always surrounds us. Through prayer, he harnessed God's mighty power one final time — and the results were clearly visible in the huge changes that occurred.

JOURNAL   My Science Notes

SPINNING SPOONS

FOCUS Weather

OBJECTIVE To explore an important weather instrument

OVERVIEW Wind speed is very important to weather forecasters. To measure wind speed, meteorologists use an anemometer. In this activity, we'll build a simple model.

WHAT TO DO

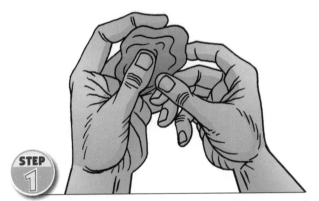

STEP 1

Discuss some ways you can tell the wind is blowing. **Make notes** about your ideas. Now **pick up** a chunk of clay and **form** it into a small cube. Keep it as square as you can.

STEP 2

Push the closed end of the syringe case into one side of the clay. The syringe case is the bottom of your anemometer. **Observe** the clay and syringe case and **make notes** about what you see.

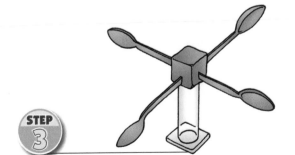

STEP 3

Push the handle of a spoon into one side of the clay. (See illustration.) **Repeat** with the other three spoons. Make sure they're spaced evenly and all face the same way. **Examine** what you've constructed so far. **Make notes** about what you see.

STEP 4

Slip a sharpened pencil into the opening of the syringe case. **Check** to make sure your anemometer can turn easily. Now **blow** gently into one of the spoons. **Observe** what happens. **Share** and **compare** observations with your research team.

WHAT HAPPENED?

An **anemometer** is a spinning device designed to measure **wind speed**. Each air cup (a spoon in your model) catches the passing air and changes it into a circular motion. In simple versions, you can get a rough idea of wind speed by watching how fast the anemometer spins. More sophisticated versions actually measure the speed of the wind in tiny increments, making them amazingly accurate.

Notice that the anemometer changes the **direction** of the **force** coming toward it. Straight wind (**linear motion**) is changed by the cups into spinning (**rotary motion**). This provides the **torque** (twisting force) that is necessary in order to take accurate measurements.

WHAT WE LEARNED

What is the name of the device used to measure wind speed? Describe how it works.

_____

_____

_____

_____

_____

_____

Describe your ideas for measuring wind from Step 1. How were they similar to an anemometer? How were they different?

_____

_____

_____

_____

_____

_____

_____

3 Describe the purpose of the clay, the syringe case, and the spoons in this model. What was the purpose of each?

4 Why is careful spoon placement important in Step 3? What might occur if they were off balance or pointing in different directions?

5 Would the model you built provide an accurate measurement of wind speed? Why or why not?

CONCLUSION

Wind has force and helps transfer energy around the Earth. Wind speed can be measured using an anemometer.

FOOD FOR THOUGHT

Mark 4:35-41 Wind is often gentle and cooling, but sometimes it can be wild and destructive! Knowing the wind speeds across broad areas helps forecasters predict tomorrow's weather. Yet even the greatest forecaster can't change the weather!

This Scripture tells how a great storm terrified the disciples and threatened to sink their boat. But when they woke Jesus, he commanded the storm to stop. Suddenly, all was peaceful again! Remember, the power that calmed that raging storm is the same power that can calm our hearts. Look to Jesus, and you can find peace.

 **JOURNAL** My Science Notes

POROUS PUMICE

FOCUS Geology

OBJECTIVE To explore properties of igneous rock

OVERVIEW Everyone knows that rocks "sink like a rock" — or do they? This is another activity that teaches us that things are not always as simple as they seem!

WHAT TO DO

STEP 1

Pour some whipping cream into a bowl. **Pick up** the whisk and begin whipping the cream. **Whip** the cream until something starts to happen. **Make notes** in your journal about what you see.

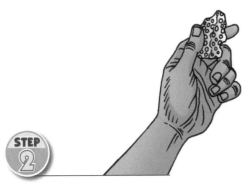

STEP 2

Carefully **observe** the piece of pumice stone. **Make notes** about similarities and differences to other rocks that you are familiar with. Now **predict** what will happen if you drop this rock into water.

STEP 3

Fill a bowl with water. Gently **place** the pumice stone in the water. **Observe** what happens.

STEP 4

Compare the whipped cream you made in Step 1 with the pumice stone. How are they similar. How are they different. **Make notes** on your journal page, then **share** and **compare** observations with your research team.

WHAT HAPPENED?

Like whipped cream, your **pumice** contains a lot of air. Since air is less dense than water (and there's so much trapped) you end up with a floating rock! Other kinds of rocks are much more **dense**, so they're the ones that "sink like a rock."

Pumice is one type of **igneous rock** (melted rock from volcanoes). Imagine a bottle of soda pop. As long as it's under **pressure**, there are no bubbles. But shake it up, open the cap — and as pressure falls bubbles shoot everywhere!

Pumice is created in a similar fashion. Deep underground there is enormous pressure. During a volcanic eruption, **magma** (melted rock) is forced to the surface and violently ejected. The instant drop in pressure results in violent expansion of **volcanic gasses**. The combination of these forces creates bubbles in the liquid rock. As the rock cools, these bubbles become trapped inside, creating pumice.

WHAT WE LEARNED

 Describe the cream in Step 1 before and after whipping. How did it look similar? How did it look different? What caused the change?

 What did you predict in Step 2? How did this prediction reflect what actually happened in Step 3?

Compare the whipped cream and the pumice. How were they similar? How were they different?

_____

_____

_____

_____

_____

Explain how pumice was made. What do scientists call the melted rock inside the Earth?

_____

_____

_____

_____

_____

Why does pumice float? How does density relate to buoyancy?

_____

_____

_____

_____

_____

? CONCLUSION

Igneous rocks are formed by melting, then hardening. Some igneous rocks, like pumice, are filled with spaces caused by volcanic gasses.

FOOD FOR THOUGHT

Romans 8:38-39 It's easy to separate pumice from other kinds of rocks. Just drop them all in the water! Other more serious separations happen too easily in this world. Circumstances may separate you from your friends, your possessions, or even your family.

Yet this Scripture reminds us that absolutely nothing can separate us from God's love! No force or power, not the devil nor the angels, not the highest mountain nor the deepest ocean . . . not even death itself can separate us from God's endless love — the most powerful force in the universe! Be sure to take time today to thank God for his amazing love.

JOURNAL My Science Notes

_____

_____

_____

_____

_____

_____

_____

_____

_____

_____

_____

_____

_____

_____

_____

_____

_____

_____

_____

_____

_____

_____

LEAKY LIMESTONE

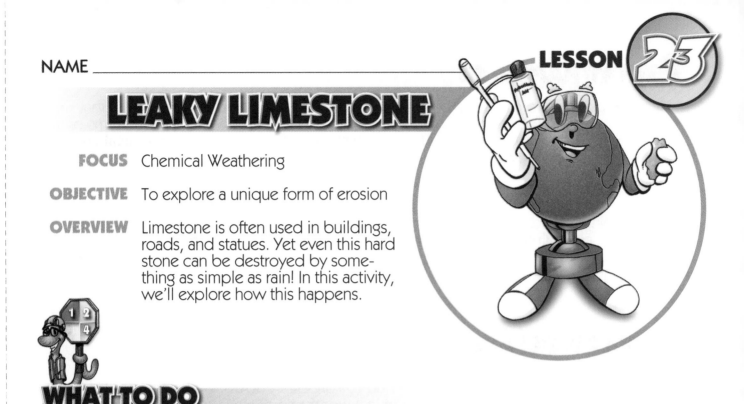

FOCUS Chemical Weathering

OBJECTIVE To explore a unique form of erosion

OVERVIEW Limestone is often used in buildings, roads, and statues. Yet even this hard stone can be destroyed by something as simple as rain! In this activity, we'll explore how this happens.

WHAT TO DO

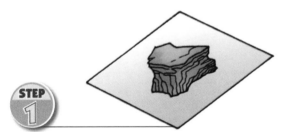

STEP 1

Closely **examine** a piece of limestone. With your research team, **discuss** some ways you might break this rock down into smaller pieces. **Record** these ideas in your journal.

STEP 2

Examine the items in your materials kit. Could any of these items have an effect on a piece of limestone? **Make notes** about why or why not.

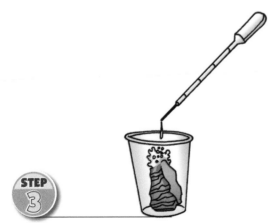

STEP 3

Place the limestone in a plastic cup. Carefully **add** hydrochloric acid one drop at a time. **Make notes** about how the liquid is affecting the rock. (Keep paper towels handy in case of a mess.)

STEP 4

Carefully **clean** your work area following instructions from your teacher. Now **review** each step in this activity. **Make notes** about any adjustments in your thinking. **Share** and **compare** observations with your research team.

WHAT HAPPENED?

Erosion is a process that breaks things down over time. There are many different kinds of erosion. For example, you've probably seen ditches or gullies that streams or rain water have cut through soil. The erosion **modeled** in this activity is **chemical erosion**. It's the breakdown of material through a chemical process.

In Step 3, the bubbling came from **carbon dioxide** released by a chemical reaction between the hydrochloric acid and calcium carbonate (limestone). So are people going around pouring hydrochloric acid on statues? In a way, that's what is happening. The **air pollution** people produce reacts with **water** in the **atmosphere** to produce **acid rain**. While the acid in this rain is too **dilute** for you to feel, over time it can create serious problems in the **environment** — not only for rivers and streams, but even for buildings or statues!

WHAT WE LEARNED

 Describe the ideas you had for breaking down the limestone. How were they similar to chemical weathering? How were they different?

 What items in your materials kit did you think might help break down the limestone? How?

3 Describe what happened in Step 3. What caused the bubbles? What three chemical substances were involved?

4 Name at least two types of erosion. How are they similar? How are they different?

5 Why is acid rain a problem? What steps might be taken to eliminate acid rain?

 CONCLUSION

There are many kinds of erosion. Chemical weathering is a form of erosion caused by acid rain.

FOOD FOR THOUGHT

Psalm 46:1-3 Although we don't often see big changes in our short lifetimes, nothing on this Earth will last forever. Earthquakes, volcanic eruptions, storms (and even little things like acid rain and erosion) continue to take their toll. Eventually, everything on Earth will wear away and be destroyed.

Scripture tells us that God's children don't need to worry — even if the world blows up and the mountains all fall into the sea! That goes for our day-to-day living too. Like acid rain on limestone, there are many little things that will erode our faith over time if we ignore them. But spending time with God each day will make us strong and keep us safe in his loving arms.

 JOURNAL   My Science Notes

FUNNEL FILTER

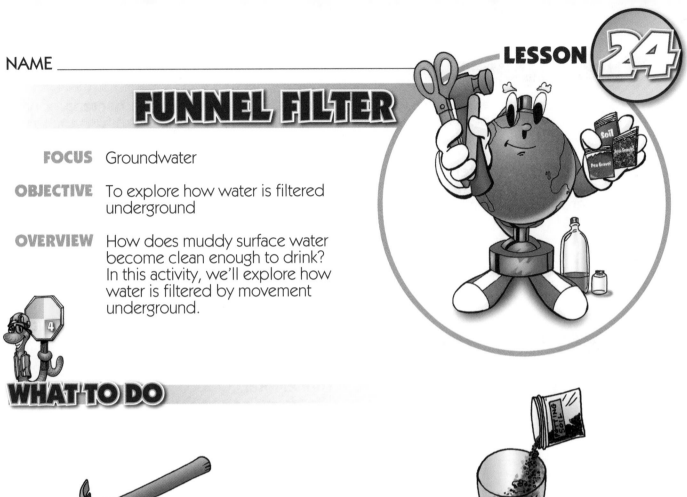

FOCUS Groundwater

OBJECTIVE To explore how water is filtered underground

OVERVIEW How does muddy surface water become clean enough to drink? In this activity, we'll explore how water is filtered by movement underground.

WHAT TO DO

STEP 1

Cut the top third off a two-liter bottle to make a funnel. (**Save** the bottom for the next step.) **Remove** the lid from the funnel. Carefully **punch** five holes in the lid with a nail. Now firmly **fasten** the lid back on your funnel.

STEP 2

Place the neck of the funnel into the bottom you saved from Step 1. **Add** one inch of pea gravel, making sure it's level. Now **add** an inch of aquarium gravel, then an inch of sand, then an inch of potting soil. **Tap** the sides several times to help the layers settle firmly in place.

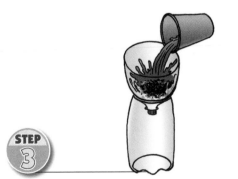

STEP 3

Stir potting soil into a cup of water until it looks muddy. **Predict** what might happen if you pour this water into your funnel. Now slowly **pour** the dirty water into the funnel. **Allow** some time for the water to trickle its way through the layers.

STEP 4

Examine the water dripping into the bottle. How does it compare to the water your poured into the funnel? **Make notes** about what you see. **Pour** the filtered water into a cup and run it through the funnel a second time. **Share** and **compare** observations with your research team.

WHAT HAPPENED?

Although this water still isn't clean enough to drink (there may be **bacteria** or other harmful **microbes**), your **Funnel Filter** did a pretty good job of removing the largest particles!

In the Earth's **water cycle**, soil, sand, and rocks perform a similar function on a much bigger scale. **Surface water** trickles down deep into the ground (now it's called **groundwater**) and **impurities** are **filtered** out in the process. For thousands of years, the Earth has been renewing water supplies this way.

However, don't assume that any clear, cold spring water is safe to drink! Modern man has introduced so many new **pollutants** into the **environment**, almost all water now requires special filters and treatments to make it safe to drink.

WHAT WE LEARNED

 Describe the layers in your Funnel Filter.
How do the particle sizes differ from layer to layer?

 What did you predict in Step 3? How did this prediction reflect what actually happened?

3 Compare the water you poured in your Funnel Filter in Step 3 with the water dripping out in Step 4. What caused the difference?

_____

_____

_____

_____

_____

_____

4 Early Americans often settled near abundant supplies of clear spring water. Is this usually a safe source for drinking water today? Why or why not?

_____

_____

_____

_____

_____

_____

5 Why is it important to control dump sites? How might something dumped on the ground miles away affect the water we drink?

_____

_____

_____

_____

_____

CONCLUSION

Earth works like a giant filter to clean groundwater. Keeping these groundwater supplies pure helps provide safe drinking water.

FOOD FOR THOUGHT

Psalm 1:1-3 Pure, clean water is vital to all living things. Water is so important to human survival that it's often used as an illustration of our spiritual and emotional well-being.

In this Scripture, the psalmist compares those who follow God to a tree planted by a stream. Because it is so close, it always has an abundant source of water. The same is true in our lives. As long as we stay close to Jesus, spending time learning to trust him each day, we'll always have an abundant source of "living" water!

 **JOURNAL** My Science Notes

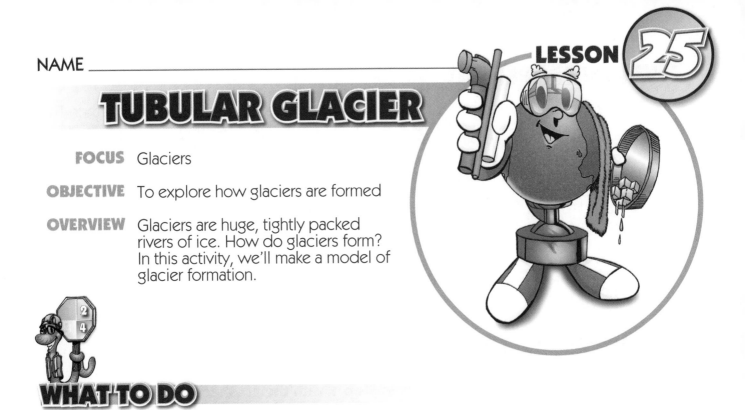

TUBULAR GLACIER

FOCUS Glaciers

OBJECTIVE To explore how glaciers are formed

OVERVIEW Glaciers are huge, tightly packed rivers of ice. How do glaciers form? In this activity, we'll make a model of glacier formation.

WHAT TO DO

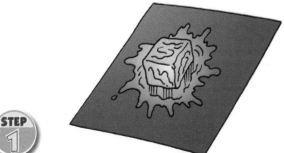

STEP 1

Carefully **examine** the ice cubes. **Make notes** describing what you see, what you feel, etc. (Science note: Ice cubes are somewhat similar, but not exactly the same as ice found deep inside a glacier.)

STEP 2

Wrap several ice cubes in an old towel. **Set** the towel on a concrete surface and **beat** it with a hammer to crush the ice. (Option: You can also use snow if there's any around.) **Make notes** about the crushed ice, comparing it to the ice cubes in Step 1.

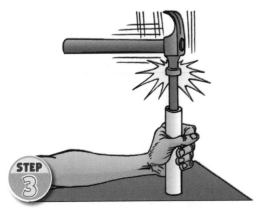

STEP 3

Pour the crushed ice (or snow) into the PVC pipe. **Pack** it as tight as you can! **Ask** a team member to hold the pipe with one end against the concrete. **Push** the wooden dowel into the end of the pipe, and carefully begin to **hammer** it down.

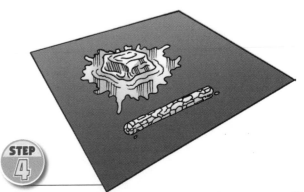

STEP 4

Once you've compressed the ice as much as you can, **lift** the pipe from the concrete and use the dowel and hammer to **tap** the "core" out the bottom. **Examine** the tubular glacier you've made. **Make notes** about what you see. **Share** and **compare** observations with your research team.

WHAT HAPPENED?

The **pressure** created by the hammer **compressed** the loose ice tightly together. This imitated the creation of **glaciers**, **which** are formed by enormous **weight** compressing snow into ice. Most glaciers begin high in mountain valleys. Falling snow piles up each winter, but doesn't all melt in the summer. Year by year, the snow pile grows higher and heavier, building up so much pressure that snow begins to compress into ice. Over time, thousands of tons of ice are created. As this massive **ice mass** (now called a glacier) continues to push down, the pressure causes the bottom layers to try to move, making the glacier slowly creep downhill.

Depending on global climate, glaciers may grow or shrink, changing the surface of the Earth beneath them. Some modern glaciers are over a mile thick! Scientists tell us the enormous weight of a huge glacier can actually push the **crust** of the Earth down several feet.

WHAT WE LEARNED

 Compare the ice cubes in Step 1 with the crushed ice or snow in Step 2. How are they similar? How are they different?

_____

_____

_____

_____

_____

_____

 Describe the process by which most glaciers are formed.

_____

_____

_____

_____

_____

_____

_____

3 How was the hammer and pipe process we used in this activity similar to the process that creates a glacier? How was it different?

4 Compare the "glacier core" from Step 4 with the crushed ice or snow from Step 2. How are they similar? How are they different?

5 Core samples from glaciers sometimes show pollution deep within the ice. Based on what you've learned about glaciers, how is this possible?

❓ CONCLUSION

Glaciers are formed when falling snow piles up year after year, creating enormous weight that compresses snow into layers of ice. A glacier's incredible weight also causes it to move slowly downhill.

📖 FOOD FOR THOUGHT

2 Corinthians 1:8-9 As you put the crushed ice or snow under great pressure with your hammer, changes occurred. The soft, pliable ice or snow became a chunk of solid ice!

In this Scripture, Paul writes about a time of tremendous physical, spiritual, and mental pressure. He and Timothy felt crushed and overwhelmed by the obstacles and problems they faced — until they put their whole trust in God! Sometimes God lets us experience hard times to help us learn to rely completely on him. Like the snow, the right kind of pressure only makes us stronger!

📖 JOURNAL   My Science Notes

NAME _____

CD SATURN

FOCUS Planets

OBJECTIVE To make a model of a planet

OVERVIEW Scientists use images (often pictures or models) to better understand things. In this activity, we'll explore a planet by making a model of Saturn.

WHAT TO DO

STEP 1

Using the Internet (NASA website, etc.), a science magazine, or an encyclopedia, **locate** a picture of the planet Saturn. **Examine** the picture closely. **Make notes** in your journal about what you see.

STEP 2

Using a sharp knife, **cut** the styrofoam ball exactly in half. **Glue** the CD to one half of the ball. Now **glue** the other half of the ball to the CD. When you're finished, the CD should be exactly centered between the two halves.

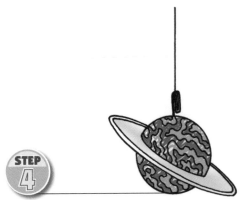

STEP 3

Using the picture of Saturn from Step 1 as a guideline, use magic markers or paint to **decorate** the planet you created in Step 2. (Hint: Pushing a toothpick into the ball will give you more control as you work.) **Glue** black yarn on the CD to simulate the rings.

STEP 4

Add glitter to simulate small pieces of debris (space junk, rocks, etc.) that make the rings. If you wish, you can **hang** your planet using a paper clip and string. Now **review** each step in this activity. **Share** and **compare** observations (and planets) with your research team.

WHAT HAPPENED?

The first hint **Saturn** wasn't simply a spherical **planet** (like **Earth**) came from a discovery made by **Galileo** in 1610. The first **telescopes** were very weak. Galileo modified the design, creating a powerful telescope that allowed him to see farther and clearer. Since then, scientists have discovered many amazing things about planets!

Your **model** resembles the view early **astronomers** might have had of Saturn. The "**rings**" they thought were **solid** (like your CD) are actually made from millions of pieces of broken rock and frozen debris. The rings only look solid because the Sun's light **reflects** from the particles' surfaces. Saturn's surface is extremely stormy, often with winds of over a thousand miles per hour, and its **atmosphere** is toxic to life as we know it. Even with a very inexpensive telescope, Saturn is fun to watch in the night sky.

WHAT WE LEARNED

 Describe the model you made in this activity. How is it similar to Saturn? How is it different?

_____

_____

_____

_____

_____

_____

 Who first discovered that Saturn wasn't similar to Earth? What device helped him in this discovery, and where did it come from?

_____

_____

_____

_____

_____

3 What are the rings of Saturn made from?
Why did early astronomers think the rings were solid?

_____

_____

_____

_____

_____

_____

4 How did the invention of telescopes help scientific discovery? Name
other "scope" devices with lenses, and tell what they're used for.

_____

_____

_____

_____

_____

_____

5 Based on present technology, could astronauts visit the surface of
Saturn? Why or why not?

_____

_____

_____

_____

_____

CONCLUSION

Each planet in our solar system has special and unique features. Devices like telescopes help scientists expand our knowledge through new discoveries.

FOOD FOR THOUGHT

Genesis 1:14-15 Your model was designed to help you better understand Saturn. Saturn is only one planet in a universe full of wonders! Step outside on a clear night, and you can see thousands of stars. Scripture tells us God created these wonderful lights in the sky. When we think of the millions of stars, planets, moons, comets, constellations, and galaxies, our minds are overwhelmed at the immensity of God's work! Isn't it amazing that despite the vastness of God's mighty creation, he still knows each one of us personally? Remember, God not only knows you, but also loves you with an endless love!

JOURNAL   My Science Notes

LESSON 27

APPLE EARTH

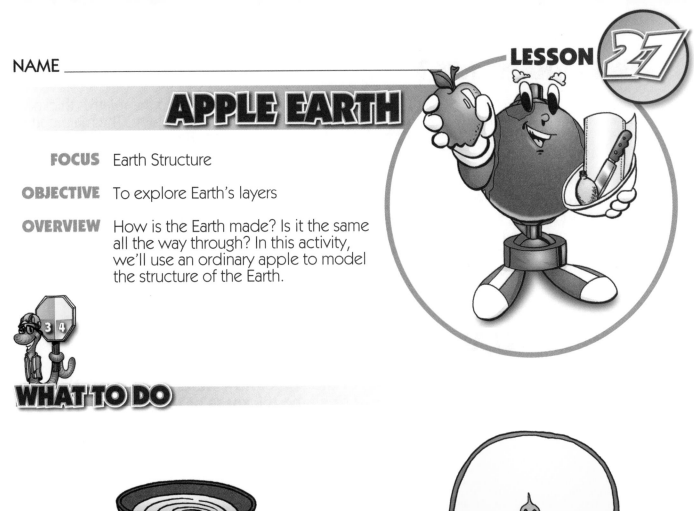

FOCUS   Earth Structure

OBJECTIVE   To explore Earth's layers

OVERVIEW   How is the Earth made? Is it the same all the way through? In this activity, we'll use an ordinary apple to model the structure of the Earth.

WHAT TO DO

STEP 1

Pour lemon juice into a bowl. Following instructions from your teacher, **cut** an apple in half. **Dip** the cut side of one half in the lemon juice. **Leave** the other half untreated. **Wait** 5 minutes, then **compare** the two halves. **Make notes** about what you observe.

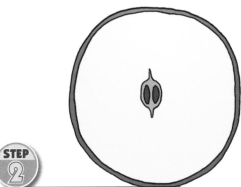

STEP 2

The parts of an apple compare well to the layers that make up Earth! **Look** closely at the circular core where the seeds are located. Now **look** closely at the "eating" part of the apple (between the core and the skin). **Make notes** comparing the two parts.

STEP 3

Look closely at the outside layer of the apple — the skin. **Make notes** describing the skin. Be sure to **compare** its thickness with the two parts you examined in Step 2.

STEP 4

Read "What Happened" (next page) to better understand how an apple's layers represent the layers of Earth. Now **review** each step in this activity, and **write** additional notes to reflect your increased understanding. **Share** and **compare** observations with your research team.

WHAT HAPPENED?

The thickness of the apple parts provides a good comparison to the relative thickness of **Earth's layers**. The core of the apple represents the **core** of Earth. The main part of the apple (between the core and skin) represents the **mantle** of Earth. The apple skin represents the thinnest layer, the **crust** of Earth.

Scientists believe that Earth's core is mostly molten iron. The mantle is made of **minerals** in a state between **liquid** and **solid** that scientists refer to as **plastic**. The crust (the part you walk on) is mostly made of **rock, soil, sand**, and similar materials. Compared to the other layers, it's relatively solid. But the crust is **brittle** and subject to violent **forces** like **earthquakes** and **volcanoes**, as well as simple (but still active) forces like **erosion**.

WHAT WE LEARNED

 1 Compare the three parts of the apple. How were they similar? How were they different? How does the thickness of the parts compare?

_____

_____

_____

_____

_____

_____

2 Using the apple as a model, name and describe the innermost (center) layer of Earth. What is it composed of?

_____

_____

_____

_____

_____

**Name and describe the middle layer of Earth.
What is it composed of?**

_____

_____

_____

_____

_____

_____

**Name and describe the outer layer of Earth.
What is it composed of?**

_____

_____

_____

_____

_____

_____

**What kind of forces affect the crust of Earth? Why do these forces
rarely affect the core or mantle?**

_____

_____

_____

_____

_____

_____

CONCLUSION

The Earth is composed of three main layers: the core, the mantle, and the crust. The relative thickness of these layers is similar to the parts of an apple.

FOOD FOR THOUGHT

2 Thessalonians 3:6-10 Models are useful in helping us understand the world around us. In this Scripture, Paul writes about the importance of being a good model to others.

Think about the way you treat those around you. Are you being a good model? Does the way you relate to others reflect the love of God? The more time you spend learning about God, the more you'll be able to truly reflect God's character to the world.

JOURNAL My Science Notes

Energy and Matter

In this section, you'll learn about the different **states** and unique **properties** of **matter**. You'll discover new things about light and sound. You'll explore physical and chemical reactions. Some grades will explore related concepts like circuits, currents, and convection.

NAME _____

LIGHT SLICER

FOCUS Light

OBJECTIVE To explore light waves and color

OVERVIEW Where do rainbows get their color? In this activity, we'll use our "Light Slicer" glasses to find out!

WHAT TO DO

STEP 1

Look around the room to see if you can see any rainbows. **Make notes** about your observations.

STEP 2

Close your eyes and **put on** your Light Slicer glasses. Now **open** your eyes and **look** around the room. **Remove** your glasses and **make notes** in your journal about what you observed. Make sure everyone in your group has a turn.

STEP 3

Put on your Light Slicer glasses again. **Look** around the room. **Remove** your glasses and use crayons or colored pencils to **draw** what you see. Make sure everyone in your group has a turn.

STEP 4

Review each step in this activity. How were the steps different. How were they the same. **Share** and **compare** observations with your research team.

WHAT HAPPENED?

Light travels in **waves**. Waves come in different lengths (called **wave-lengths**). Like a **prism**, your Light Slicer glasses can **bend** light. Bending light in this fashion separates it into different wavelengths in a process known as **refraction**. Our eyes see these different wavelengths as different **colors**.

Rainbows make colors the same way because water droplets act like miniature prisms. Do you know the colors of the rainbow in order? A great way to remember them is with the name "Roy G. Biv." That's an acronym for Red, Orange, Yellow, Green, Blue, Indigo, and Violet.

There are many kinds of waves in our world — like **light** waves, **sound** waves, and **heat** waves — but we'll save that for other lessons!

 **WHAT WE LEARNED**

1 Compare what you saw in Step 1 with what you saw in Step 2. How were the views similar? How were they different?

_____

_____

_____

_____

_____

_____

2 Make a list of the colors you saw in Step 2. What's an easy way to remember the colors of the rainbow in order?

_____

_____

_____

_____

_____

_____

3 Explain how your "Light Slicer" glasses change the way your eyes see light. What is the process called?

_____

_____

_____

_____

_____

_____

4 Explain how a rainbow creates colors.

_____

_____

_____

_____

_____

_____

5 Other than light, name as least three other kinds of waves. (Hint: What kitchen appliance uses a special kind of wave to heat food quickly?)

_____

_____

_____

_____

_____

 # CONCLUSION

Prisms can bend light, separating it into different wavelengths. Visible light has several wavelengths that humans perceive as colors.

 # FOOD FOR THOUGHT

Genesis 9:13 In this activity, you had an opportunity to explore light and see rainbows. A rainbow is one of the more amazing forms of light on Earth.

But rainbows are more than just a pretty sky coloring. Every rainbow represents a promise from God. Scripture tells us that after the Flood, God promised to never again cover the Earth with water. God placed the rainbow in the sky as a sign of this promise. And unlike people, God never makes a promise he doesn't intend to keep!

 # JOURNAL My Science Notes

PAPER MICROSCOPE

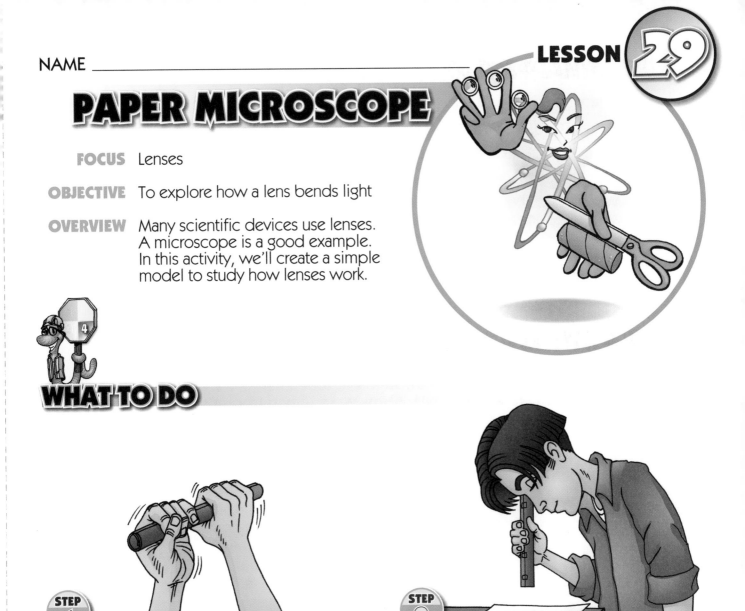

FOCUS Lenses

OBJECTIVE To explore how a lens bends light

OVERVIEW Many scientific devices use lenses.
A microscope is a good example.
In this activity, we'll create a simple
model to study how lenses work.

WHAT TO DO

STEP 1

Using scissors, **cut** down the length of the cardboard tube.
Carefully **roll** the tube tightly around two lenses (one on
each end). **Tape** the tube securely, then **pop** the lens back
out. Now, **print** your name on a piece of scratch paper.

STEP 2

Insert one lens by placing it on your work surface and slipping the tube down over it. (If it falls out, tighten the tube
and retape.) Put your eye to the open end of the tube and
look straight down at your name. **Move** your head (or the
tube) up or down to focus.

STEP 3

When your name is in focus, **freeze**! **Ask** a team member
to measure from the end of the tube to the paper. **Record**
this in your journal. Now **push** a second lens into the tube
(as in Step 2) so that it touches the first lens. **Find** the new
focusing distance, **measure**, and **record**.

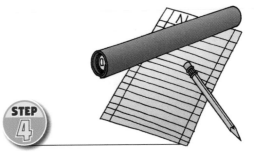

STEP 4

Review each step in this activity. **Make notes** about similarities and differences. **Share** and **compare** observations
with your research team.

WHAT HAPPENED?

A **lens** is a piece of curved glass or plastic that causes **light** to **bend** as it passes through. Some of the earliest-known lenses were made by glass craftsmen in Venice. Some of these "spectacles" were made over 650 years ago! Because people thought these little glass disks resembled lentils (a kind of legume), they became known as "lentils of glass." The Latin word for this phrase is "lenses."

Moving a lens back and forth allows you to adjust (**focus**) the amount of light hitting a particular spot and the degree it bends. This focusing makes an **image** easier to see. Today, lenses are used in many devices — microscopes, eye glasses, contact lenses, cameras, telescopes, binoculars — the list is almost endless! But every lens works on the same principle of bending light.

WHAT WE LEARNED

 Compare the views through the tube in Step 1 and Step 2. How were they similar? How were they different?

_____

_____

_____

_____

_____

_____

 Describe the surface of a lens. What happens to light when it goes through a lens?

_____

_____

_____

_____

_____

_____

3 What can we do to improve the image seen through a lens? What is this called?

_____

_____

_____

_____

_____

_____

_____

4 Name three common devices that use lenses and tell what each is used for.

_____

_____

_____

_____

_____

_____

_____

_____

5 Explain some of the history of lenses. When were the first lens made? Where were they made? Where did they get their name?

_____

_____

_____

_____

_____

_____

CONCLUSION

A lens is curved piece of glass or plastic used to bend light. Focus means adjusting the distance between a lens and an object to get a clearer view of an image.

FOOD FOR THOUGHT

Psalms 119:105 Your microscope model provided a simple way to make things easier to see. Real microscopes help scientists get a clearer understanding of how things work, and discover the answers to important questions.

While scientific tools are great for some kinds of questions, they can't provide answers to our every personal question. That's why God has given us the Scriptures! As we spend time each day reading or listening to God's Word, we discover new insights that show us the right path and keep us from stumbling.

JOURNAL   My Science Notes

NAME _____

BURBLING BIRD

FOCUS Sound

OBJECTIVE To explore how sound is made

OVERVIEW Sound is all around us. But how is sound produced? In this lesson, we'll explore sound through four simple activities.

WHAT TO DO

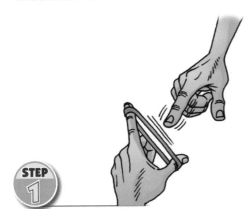

STEP 1

Fill a cup half full of water. Gently **tap** the side of the cup and **watch** the water's surface. **Make notes** about what you see. Now **stretch** a rubber band and **pluck** it with your finger. **Make notes** about what you hear.

STEP 2

Close your eyes and **sit** in a chair. **Ask** a team member to walk around the chair whistling softly. With your eyes closed, try to **point** to them. Take turns sitting and whistling. **Make notes** about what happens.

STEP 3

Blow into the tail of the Burbling Bird. **Make notes** about what happens. Now **fill** the bird with water and try again. (Keep a paper towel handy for spills.) **Record** your observations. Make sure everyone in your research team has a turn.

STEP 4

Review each step in this lesson. How were the four activities similar? How were they different? **Make notes** about your conclusions in your journal. **Share** and **compare** observations with your research team.

ENERGY • MATTER **133**

WHAT HAPPENED?

Sound travels in **waves** produced by rapid back-and-forth movements that are called **vibrations**. It requires **energy** to make something vibrate. In this case, the energy was provided by your lungs. As you began to blow, the **air** in the plastic bird began to vibrate, creating sound. Adding water changed the amount of air trapped in the bird. This changed the length of the wave and how fast it vibrated. This different vibration rate created the different sound you heard.

Almost anything that creates sound works in the same way whether it's a musical instrument (buzzing reed, plucked string, banged drum), your voice (vibrating vocal chords), or even the television or radio (vibrating speaker). The combination that produces sound is always the same — energy, vibration, sound.

WHAT WE LEARNED

 How does sound travel? What are the rapid back-and-forth movements called? What is required to make this happen?

_____

_____

_____

_____

_____

_____

 Compare the two activities in Step 1. How were they similar? How were they different?

_____

_____

_____

_____

_____

_____

3 How was the activity from Step 2 similar to the activities in Step 1? How was it different?

_____

_____

_____

_____

_____

_____

4 Compare the sound from the empty bird with the sound after you added water. How were they similar? How were they different?

_____

_____

_____

_____

_____

_____

5 List three things that produce sound. Explain how vibration is involved with each.

_____

_____

_____

_____

_____

_____

CONCLUSION

It takes energy to make something vibrate. When vibration occurs, sound is the result.

FOOD FOR THOUGHT

Mark 1:35 At first, you may have had a little trouble getting the Burbling Bird to sing properly. But after a little practice, you could warble away!

Prayer is a lot like that. The more we pray, the easier it becomes. While you can pray anywhere, earnest prayer (and listening) takes time alone with God away from the world's distractions. In spite of his hectic schedule, in spite of the press of the daily crowds, Jesus took "quality time" to meditate and pray every day. We can too!

 **My Science Notes**

MARVELOUS MAGNETS

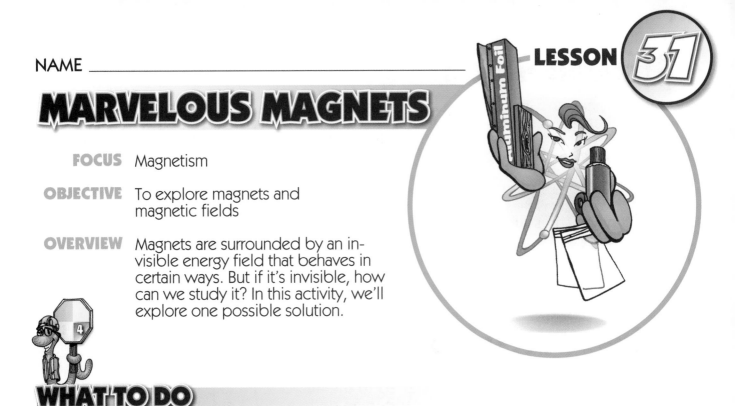

FOCUS Magnetism

OBJECTIVE To explore magnets and magnetic fields

OVERVIEW Magnets are surrounded by an invisible energy field that behaves in certain ways. But if it's invisible, how can we study it? In this activity, we'll explore one possible solution.

WHAT TO DO

STEP 1

Ask a member of another team to hold their magnet while you hold yours. **Hold** the magnets close to each other, but don't let them touch. **Turn** one magnet around and **bring** them close again. **Make notes** about what happens. (Be sure everyone on both teams has a turn.)

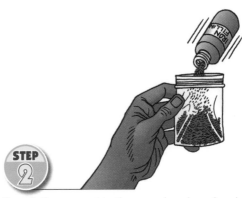

STEP 2

Touch the magnet to the wooden dowel rod, then the PVC pipe, then the aluminum foil, then the nail. **Make notes** about what happens each time. Now **pour** a tablespoon of iron filings in the plastic bag. **Leave** a little air and **seal** it tightly. **Repeat** with a second bag.

STEP 3

Stick the magnet to the center (outside!) of one bag. **Shake** the bag gently, then **hold** it with the sealed end up. **Make notes** about what happens to the iron filings. Now **place** the second bag next to the first bag with the magnet. **Shake** the bags gently and **observe** what happens.

STEP 4

Repeat the end of Step 3 until everyone has had a turn. Be sure the bags stay tightly closed so the iron filings don't touch the magnet directly! Now **review** each step in this activity. **Share** and **compare** observations with your research team.

WHAT HAPPENED?

Since iron filings are attracted by **magnetic fields**, their behavior helped you see how these fields surround a magnet. By watching the filings, we can see that magnetic fields are not flat — their influence is **multi-dimensional**. This means they affect things above or below, in front or in back, or on either side.

When you put two magnets close to each other, their behavior depended on which way you were holding them. This is because magnets have two **poles** known as **north** and **south**. Similar poles (like north and north) **push** apart or **repel** each other. Opposite poles (like north and south) **pull** together or **attract**.

WHAT WE LEARNED

 Describe what happened to the magnets in Step 1. Why did they behave differently depending on how you held them?

_____

_____

_____

_____

_____

_____

 Describe the reaction of each material in Step 2 to the magnet. What can we conclude from this?

_____

_____

_____

_____

_____

3 Based on what you discovered in Step 2, what would happen if the bag contained only sawdust or paper? Why?

_____

_____

_____

_____

_____

_____

4 Describe the magnetic field shown by the iron filings' behavior in Step 3. What additional behavior occurred in Step 4. What can we conclude from this?

_____

_____

_____

_____

_____

_____

5 What are the two ends of a magnet called? How do these two ends react to each other? What are these two reactions called?

_____

_____

_____

_____

_____

CONCLUSION

Magnets have invisible fields. Similar poles of a magnet repel each other. Opposite poles attract each other.

FOOD FOR THOUGHT

Acts 4:31 Until the magnet arrived, the iron filings were lifeless. They didn't show any movement or activity. But when the power of the magnet came near, things changed dramatically! The iron filings really became active, and they helped us see the power that was present.

Scripture tells us something special happens when God fills our lives. We may feel dull, ordinary, and powerless — but when the Spirit of God begins working in our lives, we can begin to make a real difference in the world. Like the filings made the magnet's power visible, so we can make God's power and love visible to others.

JOURNAL My Science Notes

NAME _____

STUDENT STATIC

FOCUS Static Electricity

OBJECTIVE To explore some properties of electricity

OVERVIEW Did you know your body is a source of electricity? No, a light bulb won't glow in your hand! But in this activity, we'll explore how "charged up" you can be.

WHAT TO DO

STEP 1

Inflate and **tie** the balloon. **Hold** it close to a team member's hair. **Observe** what happens. Gently **rub** his/her hair back and forth with the balloon. Now **move** the balloon an inch or two away from his/her head. **Record** the results.

STEP 2

Cut a sheet of paper into tiny pieces. **Rub** the balloon against your hair to "recharge" it. **Hold** it over the paper scraps and **observe**. Now **pour** the sawdust into a pile. **Recharge** the balloon, then **move** it close to the sawdust. **Record** the results.

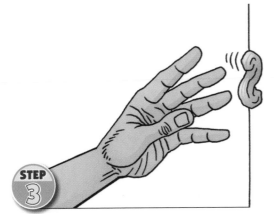

STEP 3

Recharge the balloon and **touch** it to the wall. **Release** the balloon and **observe** what happens. **Recharge** again, then **rub** a packing peanut with the balloon. **Touch** the packing peanut to the wall, then **release** it and **observe** what happens. **Record** the results.

STEP 4

Make sure everyone in your research team has a chance to try each step. **Review** each step, then **write** additional notes about what you've seen. **Share** and **compare** observations with your research team.

WHAT HAPPENED?

Rubbing the balloon back and forth captured **electrons** from your hair. Their **negative charges** (trapped on the balloon's surface) caused objects to be **attracted** to the balloon. In Step 3, you were even able to **transfer** the electrons from your hair to the balloon, then on to a packing peanut!

But just because you have an "electric personality," don't plan to light up the room! While all **electricity** is made of electrons, there are two different kinds of electricity. **Current** electricity (the kind that powers most household devices) is based on electrons that are moving. **Static** electricity (like you produced) is based on electrons that are not moving.

WHAT WE LEARNED

 Compare what happened in Step 1 before you rubbed the balloon with what happened after. How were they similar? How were they different?

_____

_____

_____

_____

_____

_____

_____

 Describe what happened in Step 2. How did the balloon affect the paper? How did it affect the sawdust?

_____

_____

_____

_____

_____

_____

_____

3 Describe what happened in Step 3. What transferred from the balloon to the peanut? What was the result?

_____

_____

_____

_____

_____

_____

4 Name the two kinds of electricity. How are they similar? How are they different?

_____

_____

_____

_____

_____

_____

5 Ten students walk across a fluffy carpet wearing wool socks. Based on what you've learned, could this generate enough electricity to light a small light bulb?

_____

_____

_____

_____

_____

CONCLUSION

Electricity is based on electrons. There are two types of electricity. Current electricity means electrons are in motion. Static electricity means electrons are stationary.

FOOD FOR THOUGHT

John 12:32 The static electricity you created in this activity has an amazing power of attraction. Whenever the balloon was close to certain objects, they were immediately drawn to it.

This Scripture talks about another powerful attraction — the pull of God's love on the hearts of his children. By demonstrating God's incredible love, Jesus drew us to him in a powerful way. As we spend time each day learning more about God's care and compassion, we will be drawn deeper into his loving embrace.

 **JOURNAL**   **My Science Notes**

NAME _____

SALTY SOLUTION

FOCUS Thermodynamics

OBJECTIVE To explore a physical change

OVERVIEW Why isn't the ocean around the North Pole frozen solid? Do solutions have different properties than water? In this activity, we'll explore how some forces interact.

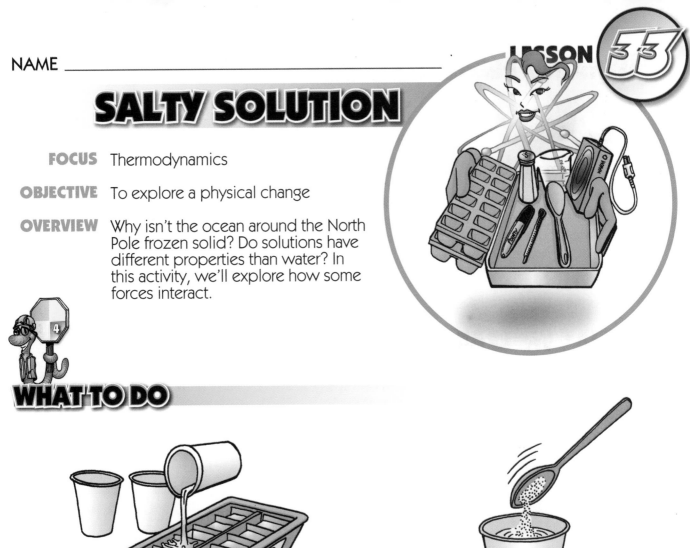

WHAT TO DO

STEP 1

Fill three paper cups with water from the sink. Carefully **empty** the cups into an ice cube tray. **Label** the tray "T" for "tap water." (Note: Masking tape makes a good temporary label.) **Observe** the water in the tray and **make notes** about what you see.

STEP 2

Refill the cups with water. Carefully **add** two heaping tablespoons of salt to each. **Stir** until the salt is completely dissolved. **Empty** the cups into a second ice cube tray and **label** it "S" for "salt water." **Observe** the water and **make notes** about what you see.

STEP 3

Place both ice cube trays in the freezer compartment of a refrigerator. **Predict** what effect if any the salt might have on the freezing point of the water. **Ask** a team member to check the trays every half hour. **Record** any changes.

STEP 4

[next day] **Remove** the two trays from the freezer. Carefully **examine** the contents and **record** the results. Now **review** each step in this activity. **Share** and **compare** observations with your research team.

WHAT HAPPENED?

The notes you recorded in your journal should indicate that the *fresh* water froze more quickly than the *salt* water. In fact, depending on how much salt you added, or how cold your freezer is, the salt solution may not have frozen at all!

Thermodynamics (thermo = heat) is the study of changes caused by **heat** and how heat moves from place to place. In this activity, **heat energy** was removed to cause a **physical change** — **freezing**. The **temperature** where a **liquid** becomes a **solid** (or vice versa) is its freezing (or **melting**) point. Liquids become solid when their **molecules** slow down so much that they stick together. The salt added to the water interfered with this process, lowering the freezing point.

Remember that freezing/thawing is a **physical change** — only the material's form changed. It's still the same material before freezing and after thawing.

WHAT WE LEARNED

 What does "thermodynamics" mean? Explain how it applies to this activity.

_____

_____

_____

_____

_____

_____

_____

 What did you predict in Step 3? How did this prediction reflect what actually happened?

_____

_____

_____

_____

_____

_____

3 Explain how liquids become solids (and vice versa). What happens to their molecules?

_____

_____

_____

_____

_____

_____

4 Did this activity represent a physical change or a chemical change? Explain your answer.

_____

_____

_____

_____

_____

_____

5 Based on what you've learned, why does adding antifreeze change the freezing point of water in a car's radiator?

_____

_____

_____

_____

_____

CONCLUSION

Thermodynamics is the study of changes caused by heat, and how heat moves from place to place. Dissolving materials in a solution can change its freezing point.

FOOD FOR THOUGHT

James 3:7-12 Looking at the water in the two ice trays, it was difficult to tell which was which before they froze. (That's why we needed the labels!) Yet even though they looked alike, the behaviors of the two were very different.

This Scripture talks about the effects of our tongue. The same mouth can say nice things, building someone up and making them feel better — or it can say mean things, hurting someone's feelings and making them feel bad. James reminds us that we need to "tame" our tongues! Before you speak, make sure the words you say are a positive reflection of the loving God you serve.

JOURNAL My Science Notes

INSTANT HEAT

FOCUS Chemical Change

OBJECTIVE To explore chemical change using a hand warmer

OVERVIEW Chemical changes always produce a different substance than the one you begin with. In this activity, we'll explore a chemical change you can use to keep your hands warm!

WHAT TO DO

STEP 1

Pick up the "instant heat" package. **Do not open** the protective wrapper! **Hold** the package in your hands. **Make notes** in your journal about its temperature. Make sure everyone on your research team has a turn.

STEP 2

Remove the inside pouch from the protective wrapper. **Shake** the pouch to activate the contents. **Wait** two minutes and **shake** the pouch again. Now **hold** the pouch in your hands. **Make notes** about its temperature. Be sure everyone gets a turn.

STEP 3

[Next day] **Allow** the pouch to cool overnight. Carefully **cut** the pouch open, and **pour** the contents onto a piece of paper. **Examine** the material and **make notes** about what you see. **Discuss** where the heat might have come from.

STEP 4

Carefully **dispose** of the pouch and contents as your teacher directs. **Wash** your hands and **clean** your work surface. Now **review** each step in this activity. **Share** and **compare** observations with your research team.

WHAT HAPPENED?

Inside this little package were **iron** filings. You couldn't see them, but you could tell by the feel of the package that they were very small. When you opened the outer wrapper, the **oxygen** in the air began to **react** with the filings in a **chemical change** called **oxidation**. This oxidation of the iron produced the **heat** you felt.

Remember, chemical change always results in a different substance. As oxidation changed the iron into a new compound (iron oxide, also known as rust), **energy** (in this case heat) was released.

Burning is another form of oxidation, only it happens much more quickly. The oxidation created in this activity is much slower than burning, but it can be very handy on a cold winter's morning!

WHAT WE LEARNED

 Compare the Instant Heat pouch in Step 1 with the same pouch after Step 2. How were they similar? How were they different?

_____

_____

_____

_____

_____

_____

 What reacted with the filings in Step 2? What is this kind of reaction called? Name another form that happens more quickly.

_____

_____

_____

_____

_____

_____

3 Describe the contents of the pouch from Step 3. What did the contents look like? Where might the heat have come from?

_____

_____

_____

_____

_____

_____

4 Did this activity demonstrate a physical or chemical change? Explain your answer.

_____

_____

_____

_____

_____

_____

5 Scientists grind a block of wood into sawdust. Later, they set fire to another block of wood. Which activity is a physical change, and which is a chemical change? Explain your answer.

_____

_____

_____

_____

_____

CONCLUSION

Chemical change always produces a different substance. Some chemical changes release energy. Oxidation is a good example of a chemical change.

FOOD FOR THOUGHT

Matthew 6:10-21   The chemical reaction in this activity produced heat, but also caused the metal to corrode rapidly. Chemical and physical changes like rusting metal, crumbling stones, and rotting wood all serve to remind us that nothing in this world is permanent.

In this Scripture, Jesus points out that everything on Earth is temporary. Although material things like our car, our home, and our clothes seem pretty important, we must never let them become more important than God! If a friend accidently stains your favorite shirt, which is more important — the shirt or your friend's feelings? Spending time with God each day helps us keep our priorities in order.

JOURNAL   My Science Notes

RUST RUSHER

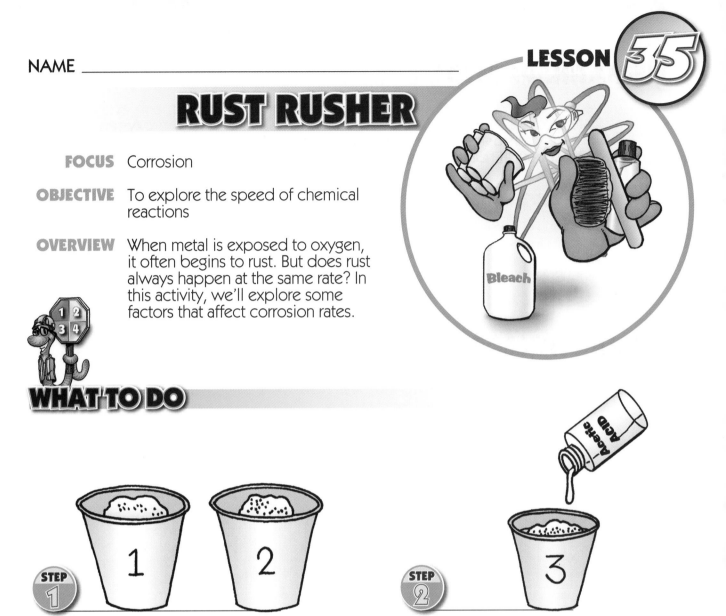

FOCUS Corrosion

OBJECTIVE To explore the speed of chemical reactions

OVERVIEW When metal is exposed to oxygen, it often begins to rust. But does rust always happen at the same rate? In this activity, we'll explore some factors that affect corrosion rates.

WHAT TO DO

STEP 1

Place a piece of steel wool in a paper cup. **Label** the cup #1. **Place** a piece of steel wool in another cup. **Fill** the cup half full of water. **Label** this cup #2. **Examine** the cups and contents and **make notes** about what you see.

STEP 2

Place a piece of steel wool in a third cup. **Fill** the cup half full of water and **add** one ounce of acetic acid. **Label** this cup #3. **Examine** the cup and contents and **make notes** about what you see.

STEP 3

Place a piece of steel wool in a fourth cup. **Fill** it half full of water and **add** one ounce of acetic acid. **Label** this cup #4. Now carefully **add** one ounce of bleach and **stir** with the craft stick. **Avoid** the fumes and don't let the liquid touch you! **Make notes** about what you see.

STEP 4

Review each step in this activity. **Examine** the steel wool in each cup and **make notes** about what you observed. **Share** and **compare** observations with your research team. **Dispose** of the materials as your teacher directs.

WHAT HAPPENED?

Even though the steel wool (which is full of **iron**) was the same in every cup, the contents of Cup #4 corroded almost immediately! When iron and **oxygen** combine, we call the **corrosion** process **rusting**. In this activity, the bleach and acetic acid created a **chemical reaction** that greatly accelerated (sped up) the corrosion process.

Although the corrosion rate varies, eventually the steel wool in every cup will rust. Many people think water causes iron to rust, but actually water only speeds up the corrosion process by keeping the iron and oxygen in constant contact.

There are entire industries devoted to protecting metal from corrosion. Galvanizing, painting, plating, and alloying are just some of the methods used.

WHAT WE LEARNED

1 Compare Cup #1 with Cup #2. How were they similar? How were they different?

_____

_____

_____

_____

_____

_____

2 Compare Cup #3 with Cup #4. How were they similar? How were they different?

_____

_____

_____

_____

_____

_____

Compare the corrosion rates of all four cups. How were they similar? How were they different?

_____

_____

_____

_____

_____

_____

When iron and oxygen combine, what is the corrosion process called? How does water accelerate this process?

_____

_____

_____

_____

_____

_____

Did this activity demonstrate a chemical or a physical change? Explain your answer.

_____

_____

_____

_____

_____

_____

CONCLUSION

The combination of iron and oxygen creates a kind of corrosion called rusting. Like other chemical reactions, the rate of the process can vary according to conditions.

FOOD FOR THOUGHT

1 Timothy 6:17-19   In this Scripture, the apostle Paul is writing to his young friend Timothy. His good advice echoes the words of Jesus that we read last week (Matthew 6: 10-21). No matter how much money someone has, no matter how many beautiful things they own, the only real treasure a person can have is what they've stored in Heaven.

How do we do this? Paul suggests that the more we do for others, the more we ourselves are changed to become like Jesus. The more we become like Jesus, reaching out to others in unselfish love, the more we invest in eternity. As someone once said, "The more you give to others, the more God gives to you!"

JOURNAL   My Science Notes

INVISIBLE INK

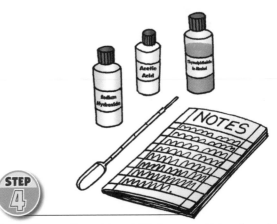

FOCUS Indicators

OBJECTIVE To explore how indicators show changes

OVERVIEW Things aren't always what they appear. Just because something is invisible doesn't mean it isn't there! In this activity, we'll learn one way to discover the unseen.

WHAT TO DO

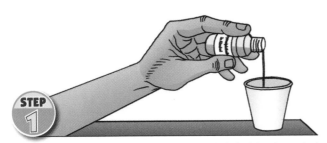

STEP 1

Pick up the bottle labeled "thymolphthalein in ethyl alcohol." **Pour** a small amount into a paper cup. Use only enough to cover the bottom — about 1/4 inch deep. **Observe** the cup and contents and **make notes** in your journal about what you see.

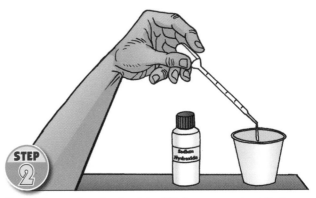

STEP 2

Open the sodium hydroxide. **Fill** the cap and set it down. (Safety note: burns!) Using the pipette, **add** a few drops from the cap to the cup. Use only enough to get a reaction. **Make note**s about what you see. **Clean up** as your teacher directs.

STEP 3

Dip the paint brush into the paper cup and **write** your name on a sheet of paper. **Blow** gently on the writing and **record** the results. Now **fill** your pipette from the paper cup and **squirt** the liquid onto a white cloth. **Make notes** about what happens. **Clean up** as directed by your teacher.

STEP 4

Open the acetic acid and **fill** the cap. Using the pipette, **add** a few drops to the cup. Use only enough to get a reaction. **Make notes** about what you see. Now **review** each step in this activity. **Share** and **compare** observations with your research team.

WHAT HAPPENED?

Thymolphthalein is a special chemical called an **indicator**. Indicators change color when **acid** or **base** levels change. Thymolphthalein is blue in a base, but clear in an acid. In Step 2, you added sodium hydroxide (a base), making a very blue solution. But when you painted this on the paper and blew on it gently, it began to disappear! This is because the **water** began to react with **carbon dioxide** in your breath to form a weak acid. Even though the **chemical change** was small, the indicator helped you discover it. In Step 4, the acetic acid neutralized the base, removing the color.

Acids and bases are measured with a system called a "**pH scale**." A reading of exactly 7 is **neutral** (neither acid nor base). Readings below 7 indicate an acid. Readings above 7 indicate a base. The pH scale is used in many ways, from environmental monitoring, to medical testing, to swimming pool tests — even checking a field's pH balance to produce better crops.

WHAT WE LEARNED

 Compare the contents of the cup in Step 1 with the contents in Step 2. How were the solutions similar? How were they different?

_____

_____

_____

_____

_____

_____

 What chemical caused the color change in Step 2? Was this chemical an acid or a base? What property of thymolphthalein helped you determine this?

_____

_____

_____

_____

_____

_____

3 Describe what happened to the paper and the cloth in Step 3. Explain why this occurred.

_____

_____

_____

_____

_____

_____

4 What are chemicals called that change color to show the presence of an acid or a base? Describe the pH scale. Tell how it is used.

_____

_____

_____

_____

_____

_____

5 Public swimming pools try to maintain pH levels around 7.5. A pH test yields a result of 6.2. Based on what you've learned, what problem does this pool have, and how might it be corrected?

_____

_____

_____

_____

_____

CONCLUSION

Indicators are chemicals that change colors to indicate an acid or a base. Indicators are measured on a pH scale. A reading of exactly 7 indicates neutral. Readings below 7 indicate an acid. Readings above 7 indicate a base.

FOOD FOR THOUGHT

Matthew 7:15-20 Indicators are a great tool for determining what something really is. A clear liquid might be cool, refreshing water — or it might be a burning acid! An indicator can keep us from harm by helping us tell them apart.

This Scripture describes an indicator you can use for people. Jesus said a person is known by their "fruit." Thorn bushes don't produce grapes, and thistles don't bear figs! Delicious fruit comes only from the right fruit tree, just as kind deeds only come from the right kind of heart (see Galatians 5:22, 23). It's only when our hearts belong to God that we can produce the right kind of spiritual fruit.

JOURNAL My Science Notes